Home Plumbing

Handyman Club Library™

Handyman Club of America
Minneapolis, Minnesota

Home Plumbing

By David Griffin

CREDITS

Mike Vail
Vice President, New Product & Business Development

Tom Carpenter
Director of Books & New Media Development

Mark Johanson
Book Products Development Manager, Editor

Dan Cary
Photo Production Coordinator

Chris Marshall
Editorial Coordinator

Steve Anderson
Senior Editorial Assistant

David Griffin
Author

Bill Nelson
Series Design, Art Direction and Production

Mark Macemon
Lead Photographer

Kim Bailey
Contributing Photographer

Bruce Kieffer
Illustrator

Craig Claeys,
Contributing Illustrator

Brad Classon, John Nadeau
Production Assistance

Dan Kennedy
Book Production Manager

ISBN 1-58159-090-3

1 2 3 4 5 6/03 02 01 00

Handyman Club of America
12301 Whitewater Drive
Minnetonka, Minnesota 55343

www.handymanclub.com

Home Plumbing

Table of Contents

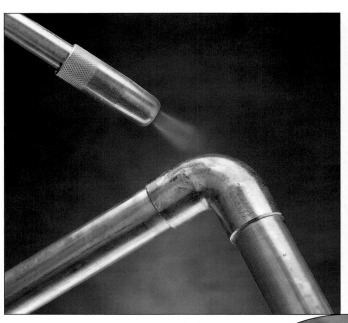

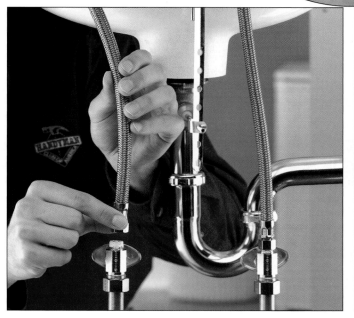

Introduction

If you own a home, you will have plumbing issues to deal with. A dripping faucet, a clogged drain, a broken fixture, a kitchen or bathroom remodeling project . . . there is just no way to avoid it. When that time arrives, you'll have two options for taking care of it: calling an expensive plumber to replace that ten-cent washer, or fixing it yourself. With the information you'll find in this book, you'll be better equipped to take the do-it-yourself route and get the job done quickly, correctly and inexpensively.

In general, plumbing is not especially complicated. It can be messy, and some of the basic skills, like sweating copper pipe, require a little practice. But if you have the desire and the know-how, along with the right tools and materials, you can accomplish most home plumbing repairs and projects.

Home Plumbing, created by and for the Members of the Handyman Club of America, is a thorough and lavishly photographed reference book designed to provide just about everything you need to know to handle most residential plumbing projects and repairs. It explains the basics of how your plumbing system works and shows how the components fit together. It identifies the pipes, fittings and other materials you'll use in your house, and demonstrates the best ways to handle them, from cutting to joining to making emergency fixes. You'll also find straightforward information and clear photographs and illustrations so you can follow along and see real plumbing projects, using actual fixtures, performed the right way. Hooking up a new toilet, installing a new sink, and replacing a bathtub or shower are just a few of the many projects presented for you. Along the way, you'll see specific examples of plumbing repairs as they're made, from emergency treatments for a burst pipe, to thawing frozen pipes, to stopping that annoying faucet drip that's been keeping you awake at night for months.

You don't need a lot of fancy, expensive equipment to arm yourself against plumbing problems. A couple of wrenches, a cheap propane torch, a hack saw or tubing cutter, and a screwdriver or two will go a long way toward meeting your plumbing tool needs. Throw in a tub of plumbers putty, a tube of pipe joint compound and roll of Teflon tape, and you'll be well prepared—as long as you also have the most important tool: knowledge.

You've probably heard plenty of home improvement horror stories (or maybe even experienced one yourself) where an unsuccessful plumbing repair, intended to fix a simple problem, leads to a major disaster that ends up costing hundreds or even thousands of dollars to have fixed. In almost all cases, these "horror stories" result from a lack of understanding of how a plumbing system works. Once again, *Home Plumbing* is just the insurance you need to prevent a predicament you'll truly regret. Take a few moments to examine this new book from The Handyman Club of America. We think you'll find it to be an indispensible reference volume you'll want to keep close at hand for the time when your next plumbing issue arises.

PLUMBING CODES

Because of the potential for disastrous water damage and the high cost of repairing a plumbing system, plumbing is a closely regulated practice. The Uniform Plumbing Code (UPC) is a national set of codes that's updated every three years. It is the foundation of most local plumbing codes. But local codes often vary from the UPC standards— usually on the more restrictive side. Use the UPC as a general guide when planning a plumbing project or repair, but make sure to consult with your local building inspection department before beginning any work.

IMPORTANT NOTICE

For your safety, caution and good judgment should be used when following instructions described in this book. Take into consideration your level of skill and the safety precautions related to the tools and materials shown. Neither the publisher, North American Membership Group, Inc. nor any of its affiliates can assume responsibility for any damage to property or persons as a result of the misuse of the information provided. Consult your local building department for information on permits, codes, regulations and laws that may apply to your project.

Typical whole-house plumbing system

Key:
- = Cold water supply
- = Hot water supply
- = Drain system
- = Drain vent system

Plumbing Basics

Plumbing is simple in principle. Water is delivered to appliances and fixtures by supply and distribution pipes. Wastewater is removed by gravity through a ventilated system of drain and waste pipes called the drain, waste, and vent (DWV) system.

Supply and distribution. A main water supply line comes into the house under pressure through a water meter and one or two shutoff valves. If you have a private well, water is pumped into a pressure tank. The pump automatically keeps the tank pressure within a set range. No meter is needed on a pressure tank, but a shut-off valve will exist on one or both sides of the pressure tank. If water quality is poor, the supply line may next pass into a water softener or a whole-house filter. Ion-exchange water softeners (See pages 155 to 157) lengthen the life and efficiency of the hot water heater, pipes, and appliances and help water work better with soaps and detergents. Whole house filters improve the smell, taste, appearance, and safety of potable water.

The cold-water distribution pipes branch as much as necessary to carry water to the fixtures (sinks, toilets, baths) and appliances (washing machines, ice makers, boilers) of the house. Soon after entering the house (or exiting the pressure tank) a line branches off to the water heater. Branching hot-water distribution pipes from the water heater then parallel the cold-water distribution pipes to appliances and fixtures in the house.

In a modern, code-compliant house, all fixtures and appliances have individual shutoff valves. These can be straight or angled, chromed, plain brass or plastic. Tub and shower shutoffs may be accessed through a panel or from below. Sinks typically have angle stops directly below the basin.

The drain/waste/vent (DWV) system. After the water is used it enters drain or waste pipes, which carry it by gravity to the sewer or septic system. Since sewage generates noxious and even explosive gasses, each point of entry into the drainage system must be blocked with a low bend of pipe filled with water. This bend is called a "trap". Every appliance and fixture that is drained must have a trap that always remains filled with water. In the early days of plumbing, traps would sometimes get suctioned out as a slug of water traveled down

The home plumbing system is a network of supply, drain/waste and vent pipes that function together to keep the water flowing in and the waste flowing out. The supply pipes are pressurized to deliver water to all levels of the home. The drain and waste pipes rely on gravity to carry waste out of the home.

drain and waste pipes, pulling the trap water along after it. Today, the J-bend of a P-trap stays full of water because air enters the drainage line after the trap through a vent, breaking the sucking action falling water in a drain line would exert on this trap water.

The toilet forms its own integral trap; the water in the bowl is the front, visible portion of the water seal. The toilet water does get siphoned during a flush, but the last water to fall from the tank automatically restores the seal. Like drain lines from other fixtures, toilet drain lines need to be vented. Often toilets are vented directly by the stack vent, which rises from a vertical soil stack.

Branch drains from fixtures on upper floors must either drop steeply or vertically to the horizontal building drain or run almost horizontally (at a ¼ in. drop per foot) to a soil stack. In-between slopes can cause clogs, since the water would rush away leaving solids stranded to snag more debris. On the slab or at basement level, stacks from upper floors enter the building drain. Fixtures and appliances on the ground or basement level empty into the building drain through vented horizontal branch lines. The building drain runs at a gentle slope to a municipal sewer or to the sewer line leading to a private septic tank.

Cleanouts are access fittings with removable caps, usually located where the drain and waste lines change direction and at the ends of drain and waste lines. Sometimes the building drain will have a house trap or "Kelly" fitting with cleanout accesses just inside or outside the house foundation. These fittings allow a mechanical auger to feed either toward the building drain or toward the sewer in the event of a clog.

Working with your plumbing system. The first step in becoming your own home plumber is to get to know the plumbing system in your house, inspecting it as you work your way from the point of entry, throughout the distribution system and back to the point of exit. Find the water meter first. Look in your basement, or a service crawl space facing the street, or possibly (in areas without hard freezes) outside. The water meter is a useful indicator of a hidden water leak. If you suspect a hidden leak, turn off valves to appliances that may draw water automatically, including icemakers, humidifiers, furnaces with humidifying units and boilers. Record usage numbers and mark the position of dial(s). Don't use water for an hour or two, then check the meter for changes before resuming water use.

Next, find the main cut-off to the water service,

which is the valve you can shut down in the event of a plumbing emergency. You should find a valve on either or both sides of your meter. These valves are apt to have drain cocks, which can be opened when the valve is shut to help drain the system. If water is billed at a fixed rate and you have no meter, or if you have a pressure tank from a private well, look for a valve near to where the water enters the house. A second valve may be located on the outlet side of a pressure tank for a well.

With a well, you should also locate a switch, circuit breaker or fuse that will allow you to cut off power to the pump. The power cut-off may be located in a disconnect box near the pressure tank or on your main electric service panel to the house. Shut off the pump when you need to shut down your system for repairs.

NOTE: A gate valve at the meter that no longer lifts the gate up all the way may cause low water pressure. If your old gate valve at the meter fails to turn off all the way, or open all the way, have the city turn off the water outside the house so you can replace your shut-off with a full-port ball valve. With municipal water service, the water company may turn off your water with a special long-handle "key" at an underground buffalo or stop box near the street.

Usually, the first stop for the water once it has entered through the meter is the water heater. Water heaters have shutoffs on the cold-water supply line to the heater and sometimes on the hot-water pipe leading from the heater. If you need to work on the hot-water supply system, you should shut these valves and also turn off the burner or heating element. Keep an eye out for signs of leaks, as you visually trace the path of the visible branch lines. Fittings and valves are more likely to develop leaks than lengths of pipe. Sometimes opening a shut-off valve all the way can stop a leak from the valve stem by compressing the packing under the packing nut. A drain line may conceal leaks from the hot water temperature and pressure relief valve located high on the side or on the top of the water heater. Feel the relief valve line six feet away from the valve to determine if hot water is leaking into it, or trace the drain line to its outlet to see if water is dripping.

Make sure your water pipes are well supported. Sometimes pipe hangers drop out of joists, leaving horizontal stretches of pipe sagging, which stresses the pipe's joints and may produce leaks and allow pipes to rattle. Plumbers tape, a kind of metal strapping with holes in it, can be purchased in rolls to re-secure loose pipes, and a range of plastic and insulated pipe hangers are available as well. Make sure you use hangers or tape that are suitable for your kind of water pipe. For example, galvanized pipe straps should never be

Water Conservation

Every year our water resources become more precious, as agriculture, industry, and municipalities vie for over-partitioned rivers, reservoirs and shrinking underground supplies. Many communities have been required to cut back on home water use, and the trend will only broaden in the future. Whether you're facing an emergency water shortage or are trying to reduce consumption over the long haul, the tips below can help your household conserve water.

WHERE THE WATER GOES. To prioritize your water saving strategies, it's important to understand how most of your water is used and where the greatest opportunities for conservation lie. Inside the house, toilets consume the most water on average, followed by tubs and showers, washing machines, dishwashers and kitchen sinks. Outside water use varies from home to home, but a medium-sized lawn with an uncovered swimming pool can consume more water per month than the entire inside water use of a family of four.

TIPS FOR CONSERVING WATER
• *Flush less.* Toilets use the most water inside the house, so save more by flushing less. You can also use less flush water by replacing an old 3½ or five gallon tank toilet with a new 1.6-gallon model. Or, buy a retrofit device for your old five-gallon tank. A variable buoyancy flapper lets less water out the flush valve, saving up to two gallons per flush. This device works better than decreasing the water volume in the tank because you have the full weight of a large tank of water to push out the flush water at the bottom of the tank. Another clever and effective retrofit is the dual-handle mechanism, which gives you two handles in one for two flush options. This allows you to flush liquid wastes with much less water than solid wastes.

• *Run faucets less.* A running faucet uses 300 gallons of water per hour. Washing dishes with the water running can use ten or twenty times as much water as using separate soap and rinse basins. Pre-rinse dishes for the dishwasher in a pan. Don't thaw food under a running faucet. Keep chilled water in the refrigerator so you don't have to run the water before filling your glass. Don't let the water run while brushing your teeth or shaving. Spout-tip shutoff valves can make shutting on and off the water second nature. Stream-spray aerators attached to the bathroom or kitchen faucet let you switch easily to a water-saving spray option for hand washing and dish rinsing.

used on copper pipe and vice versa.

Trace the tandem, hot-and-cold supply lines through your house, noting the location and condition of shut-off valves used to control the water flow to individual fixtures and appliances through out the home. Don't forget to follow the cold water supply lines to exterior faucets (called sillcocks). Test each shut-off valve to make sure it's in working order (better to find out now than to wait until your faucet supply tubes bursts and you're unable to stop the water flow at the shut-off). Also check the general condition of all faucets and spouts before turning your attention to the DWV system.

DWV pipes perform the critical, if inglorious, task of carrying away waste from appliances and fixtures. Try to identify some of the features of your system. If you look under your kitchen and bathroom sinks, you may find P-traps, S-traps, or pipes dropping straight to the floor with no traps visible. Don't worry, these seemingly trap-less fixtures probably drain to a drum trap beneath the floor. Sometimes a sink and a bathtub will share a drum trap. Take a moment now to locate your traps and see if you could open them if they became blocked. P-traps are sometimes assembled with slip nuts that are easy to remove, and drum traps have a removable cap, which can face up through a floor or down from a ceiling.

Try to identify components of the vent system. The attic is a good place to look for these. The stack vent (usually cast iron in older homes, more likely PVC in

Main shutoff

The water meter is the gatekeeper of the flow of water into your home. You'll need to know the location of the meter so you can shut off water service to the house in emergencies or for plumbing projects.

new houses) is usually the principal vent. Other branch vents are back-vented to this. The stack vent is a continuation of the soil stack—the big vertical pipe that drains one or more toilets and also receives branch waste lines from other fixtures and appliances. Fixtures and appliances far from the stack may have their own waste stack to the building drain and also join to a separate vent through the roof. The important point is that every plumbing fixture and appliance in your home has both a trap and a vent just beyond the trap.

You may find cleanouts near the terminus of branch lines near fixtures and appliances, or sometimes a removable trap is considered a cleanout access. You will likely find removable cleanout plugs on change-of-direction fittings on branch lines, where the stack joins the building drain and where the building drain joins the sewer. This final cleanout may be on the inside or outside of the foundation wall.

Possible problems in you DWV system include leaks, failure to drain, and failure to keep sewer gasses out of the house. Of course, drainpipes will leak only when wastewater is flowing through them. Look for stains on and near pipes that may indicate leaks. Poor drainage performance can result from clogged or inadequate vents or clogs in the drainpipes. Drainpipes should be supported with heavy-duty plumber's tape or hangers. Replace or reattach sagging hangers, or add hangers to straighten sagging runs of pipe.

• *Fix leaks.* A leaky faucet or running toilet can waste hundreds or even thousands of gallons of water a month. Keep valves in fixtures and appliances working properly according to the instructions provided in this book.

• *Run washing machines and dishwashers fully loaded.*

• *Bathe efficiently.* Turn off the water while soaping in the shower and use a water-saving showerhead. A high quality showerhead or shower massager forms and delivers water droplets so the shower experience is not compromised by lower water volume. A flow control lever between the shower arm and head lets you reduce water to a trickle while lathering. The trickle maintains the water temperature.

• *Minimize lawn watering.* One study indicated that homeowners consistently over-water their lawns by 100%. Water only when needed and where needed. Switch automatic sprinkler systems to "manual" to avoid watering during wet and cloudy periods. Keep sprinkler heads or drip emitters clean and properly directed. Surface water only in the early morning and evening. Use drip irrigation systems rather than sprinklers for gardens and landscaping plants. Check for concealed outdoor leaks.

Plumbing Tools

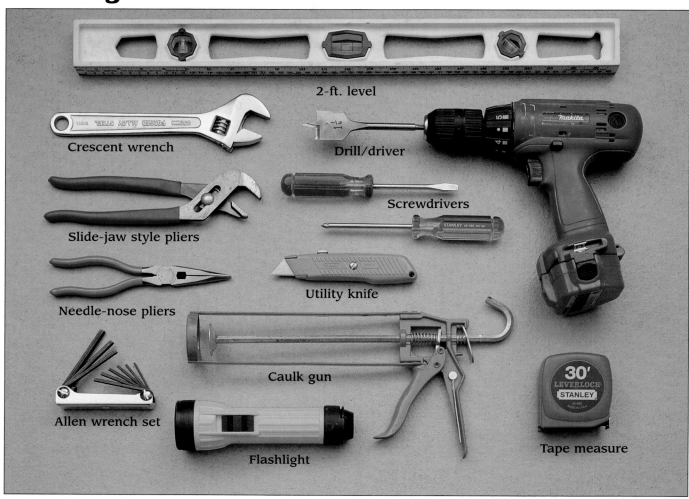

2-ft. level

Crescent wrench

Drill/driver

Screwdrivers

Slide-jaw style pliers

Utility knife

Needle-nose pliers

Caulk gun

Allen wrench set

Flashlight

Tape measure

Everyday shop tools. Before starting a plumbing project, make certain your workshop has the right stuff. These tools are common to many repair projects, but don't get stuck in the middle of a plumbing job when you discover you are missing an obvious tool.

Cutting tools. When parts of your plumbing system need repair or replacement, it's likely you're going to be doing some cutting. Choose the right tool. Your results will be better, and you'll work more safely.

Power miter box (A) is great for cutting pipe; *reamer (B)* is necessary when threading galvanized pipe; *galvanized pipe cutter (C)* makes the neat, even cuts you need when joining galvanized pipe; *plastic tubing cutter (D)* works like pruning shears to cut plastic pipe quickly; *hacksaw (E)* is the simple workhorse of many plumbers; *plastic pipe saw (F)* slices quickly through plastic pipe, especially large-diameter pipe; *reciprocating saw (G)* will cut through pipe, studs, nails or just about anything that needs removal; *pipe threader (H)* will cut threads on galvanized pipe; *copper tubing cutter (I)* cuts copper tubing neatly, even in tight areas.

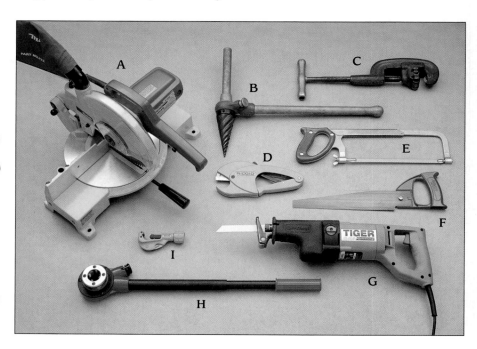

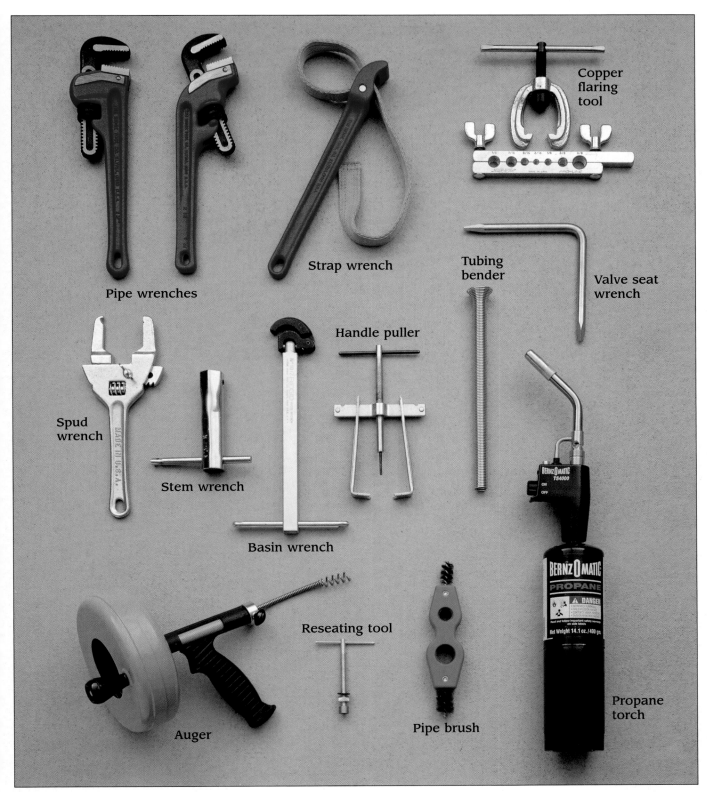

Copper flaring tool

Pipe wrenches

Strap wrench

Tubing bender

Valve seat wrench

Spud wrench

Stem wrench

Handle puller

Basin wrench

Propane torch

Auger

Reseating tool

Pipe brush

Specialty plumbing tools can help you do the job right, and generally are not too expensive. Buy them on an as-needed basis.
Pipe wrenches for turning galvanized pipe—you'll need two for most jobs; *strap (or cloth) wrench* for removing easily-marred chrome pipes; *copper flaring tool* shapes the ends of copper tubing for flare fittings; *valve seat wrench* removes the small valve seats from worn faucets; *spud wrench* tightens large nuts that are 2 to 4 in. in diameter;

stem wrench removes faucet stems hidden in recessed holes; *basin wrench* twists off hard-to-reach nuts on plumbing fixtures; *handle puller* will free faucet handles from corrosion bonds; *tubing bender* ensures neat, kink-free copper tubing bends; *auger* clears drain lines of debris; *reseating tool* will grind a smooth faucet seat; *pipe brush* cleans pipes prior to gluing or soldering; *auto-igniting propane torch* provides heat for soldering copper and other plumbing projects.

Supply pipes

Flexible copper tubing is used mainly for gas, but the narrow water supply tubes to fixtures and appliances may be made of flexible copper tubing. Cuts with a tubing cutter or hacksaw. Joins with soldered, compression, or flare fittings.

Rigid copper is the most common water distribution material. Type M (with red lettering) is the lightest type and is used most commonly above ground in residences. Joins with soldered fittings. Cuts with a copper-tubing cutter or a hacksaw. Compression fittings may be used if joints are exposed.

CPVC (Chlorinated polyvinyl chloride) is a rigid plastic tubing used in hot and cold-water distribution. CPVC is popular among DIYers since its joints can be solvent welded. Check local codes before using. CPVC is solvent-welded into fittings, or grip fittings are used. Cut with a PVC tubing cutter, hacksaw, or miter box and saw.

PEX (Cross-linked polyethylene) is a flexible plastic tubing used in hot and cold-water distribution. Check with local code officer before using. PEX joins with insert fittings and crimp rings or with grip fittings. PEX cuts with a flexible plastic tubing cutter, shears, or a knife. PEX should not be left exposed to direct or indirect sunlight indefinitely.

Galvanized steel is found in older dwellings and is susceptible to water pressure loss due to scale buildup. You may transition to another material with the appropriate transition fittings when extending water distribution lines. Joins with threaded fittings. Cuts with hacksaw, reciprocating saw, or ferrous pipe cutter.

PE (Polyethylene) is a cheap, rugged, flexible plastic pipe used for cold water. Common uses include water supply to the house, irrigation, and cold water supply to an outbuilding. PE is joined with PVC insert fittings and hose clamps. It cuts with a flexible plastic tubing cutter or a sharp knife. For irrigation, a cheaper utility grade PE tubing is used.

Plumbing Materials

Pipe and tubing are the most basic plumbing materials. The materials and their characteristics have evolved over the years in response to changing technology, availability, cost and health concerns.

You will most likely encounter one of seven kinds of water distribution pipe in your house. Identifying and measuring them correctly is critical to purchasing the right fittings and valves. Identify newer pipe by the codes printed on the pipe. The nominal diameters of copper supply and distribution pipes are about equal to the inside diameters of the pipes. The nominal diameters of copper and plastic water pipe are 1/8 in. less than the outside diameters. Small diameter supply tubes beneath fixtures and appliances are measured by their outside diameters (except synthetic and woven flexible tubes). You may switch pipe materials on a job if you use the appropriate transition fitting (See pages 16 to 18).

Take samples, measurements, and written information from the sides of the pipes when going to a plumbing supply store for fittings. The lettering on the sides of pipes says everything. Red lettering on copper indicates type M pipe, the thinnest wall thickness grade and the kind you are most likely to use. "NSF-PW" and

Standard pipe diameters		
Material	Inside dia.	Outside dia.
Copper	1/4	3/8
	3/8	1/2
	1/2	5/8
	3/4	7/8
	1	1 1/8
	1 1/4	1 3/8
	1 1/2	1 5/8
Galvanized steel	3/8	5/8
	1/2	7/8
	3/4	1
	1	1 1/4
	1 1/4	1 1/2
	1 1/2	1 3/4
	2	2 1/4
Plastic (Schedule 40 PVC)	0.5	0.845
	0.75	1.050
	1	1.315
	1.25	1.660
	1.5	1.9
	2	2.375
	3	3.5
	4	4.5
Cast iron	2	2.35
	3	3.35
	4	4.38

"NSF-DWV" mean the National Sanitation Foundation has rated the pipe for pressurized water or drain-waste-vent, respectively. The only kinds of plastic pipe currently installed for hot and cold supply and distribution are CPVC and PEX. Plastic water pipes follow the sizing convention established for copper water pipe. The outside diameter of a CTS (copper tube size) pipe is ⅛ in. greater than its nominal diameter. On copper pipe, nominal and inside diameter are about equal.

Old houses may have cast iron or terra cotta pipe joined together with flared hubs and straight spigots. These joints will be packed with hemp or oakum and sealed with lead. More recent cast iron and terra cotta pipes are "no hub." No-hub pipe and fittings are the same diameter end-to-end and are joined with stainless steel clamps over rubber sleeves. No-hub cast iron is still used extensively in modern construction because it is quiet, durable, and it won't burn.

Threaded galvanized steel pipe and soldered copper were used extensively in DWV systems for branch lines before the advent of plastic. Schedule 40 plastic DWV pipes share the same outside diameters as steel pipes of the same nominal size. Most residential drain systems now are partially or wholly made of black ABS plastic or white PVC plastic. These light, strong pipes are easily cut and cemented together using hubbed fittings.

Plumbers putty (left photo), **pipe joint compound** and **Teflon tape** (right photo) are three products you'll use extensively in residential plumbing. Generally, joint compound is applied to female threaded fittings, Teflon tape is applied to male threaded fittings and plumbers putty is used to create a waterproof seal around fixtures in wet areas (for example, between a sink rim and countertop).

DWV pipes

Galvanized steel was once (and sometimes still is) used for DWV branch lines. It is joined with threaded fittings. Galvanized cuts with a ferrous pipe cutter or reciprocating saw. It is threaded and joined with fittings or with threaded union fittings.

Copper is rarely used for DWV branch lines or traps anymore. It is joined with solder and sometimes with banded clamps over rubber sleeves. Copper may be cut with a tubing cutter, a reciprocating saw, or a hacksaw. Banded couplings can be used to join other pipes and tubing to copper drain and waste pipes.

ABS (Acrylonitrile Butadiene Styrene) plastic is used in modern plumbing. It is cheap and easy to work with but it will burn and it dampens less noise than cast iron. Cut with a tubing cutter, plastic saw, or miter box and saw. Join with ABS solvent cement.

PVC (Polyvinyl Chloride) plastic is the most commonly used DWV pipe material. It is resistant to damage by heat and chemicals. PVC is cheap and easy to work with but will burn and is loud. Cut with a tubing cutter, plastic saw, or miter box and saw. Join with PVC solvent primer and cement. Schedule 40 PVC is shown here.

Cast iron was once common for drains, particularly the main drain stack and the floor drain. The outside pipe (inside hub) diameter is from ¼ to ⅜ in. greater than the nominal (inside) diameter. Pipe marked XH is ⅜ to ½ in. bigger than nominal. Iron pipe cuts with a snap cutter and may be joined with solid banded couplings with rubber or neoprene sleeves inside.

Couplings, fittings & adapters

Couplings, fittings and adapters are used to join the various components of your plumbing system together. Whether they're copper, PVC, brass or another plumbing material type, they all look pretty much the same in shape and relative size. The examples shown on these two pages are copper and brass fittings, all of which are soldered to copper pipe on one or both ends. If you are installing or working with PVC, galvanized or another pipe material, look for fittings made for that material. The different configurations are designed to make bends in the lines, make transitions and extend pipe on longer runs.

Typically, the method of joining a fitting to pipes and other fittings reflects the same technique used in making other types of connections with that material. Exceptions to this include some types of supply tube fittings (See below) and banded couplings (See page 18). Adapters and transition fittings generally utilize the connection type used for each pipe at each end of the fitting.

Supply fittings

Compression fittings

Retainer nut

Ferrule

Union

Ferrule

Retainer nut

Compression fittings are useful where connections are not permanent. Plumbers use compression fittings mainly to attach small diameter water supply tubes to stop valves beneath sinks, lavatories, toilets and appliances. Code usually prohibits concealing compression fittings in walls, since they are more prone to failure than soldered or solvent welded joints. Do not use compression fittings where they will be subject to vibration, which could loosen the compression nut. See page 27 for more information.

Flare fittings

Flare nut

Flare union

Flare nut

Flair fittings work something like compression fittings, except instead of a brass ferrule, the flared end of the pipe itself is compressed between the fitting and the nut. Flair fittings are not used much with water supply pipes and tubes anymore, but they are still popular with copper gas and fuel lines. See page 33 for more information.

Drop eared elbow. The drop eared elbow (also called a drop-eared 90) is commonly used when working with plumbing fixtures. It lets you transition from a vertical pipe in the wall to a horizontal copper stub out, brass nipple, or shower arm. The ears let you screw the elbow to blocking, adding stability and strength to whatever attachment extends outside the wall. To the right is a ½ in. cup × FIP (female iron pipe) drop-eared elbow that transitions from copper pipe to a threaded brass nipple.

Repair coupling. When you are doing repair work or adding a branch line, it is often necessary to insert a fitting or assembly into fixed pipes. Couplings without stops (also called "repair couplings) are needed here to allow the coupling to slide out of the way onto a leg of the pipe while you insert the new pipe. Couplings without stops may be purchased for PW and DWV piping in both plastic and copper.

Threaded Adapters. Threaded adapters can be soldered to copper or solvent welded to plastic, enabling you to make a connection with threaded pipes or nipples. The threaded side of the adapter can be screwed to another threaded fitting or to a threaded pipe. Threaded adapters provide a safe, strong way to transition from copper to plastic. It is preferable (and frequently required by code) that the plastic adapter have the male end and the metallic adapter have the female end. This is because threaded pipe and fittings are tapered, and screwing a hard metal male end into a soft plastic female socket can break the female fitting apart. The Plastic MIP adapter should be wrapped clockwise with four layers of Teflon tape. Apply a pipe joint compound that is compatible with the plastic material on the metallic female threads.

Female

Male

Stop coupling. When you need to joint two straight runs of pipe in a new installation, use a stop coupling to make the connection. Unlike the smooth repair coupling, the stop coupling has internal ridges that the mouth of each tube press against when inserted, strengthening the bond when the parts are soldered or solvent-welded.

Street or male-end fittings. Sometimes you need to insert one fitting directly into another fitting or valve without any pipe in between. The inserted fitting must be "streeted"—that is, it must include a port that shares the same outside diameter as a pipe. Street fittings are common with DWV pipes, where wide pipes need to change direction in small areas. In water supply, you may need copper street elbows to join a single-handle tub and shower valve to the hot and cold water supply

Unions are a secure mechanical coupling for pipes that can be taken apart with wrenches without twisting the pipes themselves. You may use unions at water heaters, whole house filters, or in other situations where you want a joint that can be opened but that provides a higher level of security than a compression fitting. When cutting your pipe for a union, make sure to allow for the gap that will be occupied by the union fitting.

Reducing fittings. Often, the openings on an elbow, tee, or coupling will require different sized cups. Reducing fittings are available in a broad range of configurations, with tees and couplings having the greatest range of options. To the left is a ¾ × ¾ × ½ tee fitting.

Transition fittings connect two different pipe materials, such as galvanized to copper, CPVC to galvanized and CPVC to copper. A galvanized-to- copper union fitting (called a dielectric union) contains a non-metallic spacer so that the two materials don't react electrochemically and corrode.

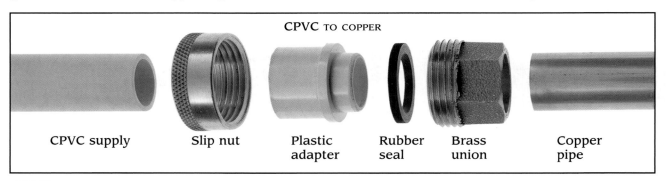

CPVC TO COPPER

CPVC supply   Slip nut   Plastic adapter   Rubber seal   Brass union   Copper pipe

CPVC to copper

When attaching CPVC to copper tubing, a four-piece union is needed. First, solder the brass union to the copper tubing. Let the brass cool before inserting the rubber seal. Then attach the plastic adapter to the CPVC pipe with solvent glue. Make certain the slip nut is in place before gluing. After the glue has set (at least 30 minutes), tighten.

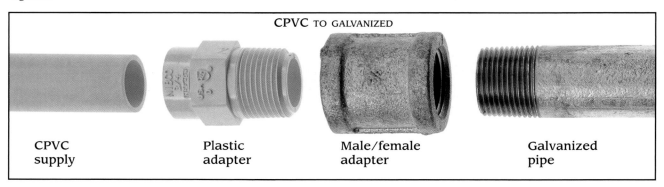

CPVC TO GALVANIZED

CPVC supply   Plastic adapter   Male/female adapter   Galvanized pipe

CPVC to galvanized

Connect galvanized pipe to CPVC pipe with male and the female threaded adapters. The plastic adapter is attached to the CPVC pipe with solvent glue. Threads of metal pipe should have a coating of pipe joint compound or teflon tape. The metal pipe is then screwed directly to the adapter.

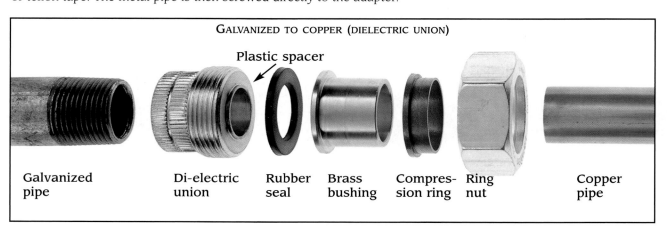

GALVANIZED TO COPPER (DIELECTRIC UNION)

Plastic spacer

Galvanized pipe   Di-electric union   Rubber seal   Brass bushing   Compression ring   Ring nut   Copper pipe

Galvanized to copper (Dielectric)

A dielectric union is needed to connect copper to galvanized steel. The union is soldered to the copper and threaded onto the steel. The dielectric union has a plastic spacer that prevents corrosion caused by an electrochemical reaction between the different metals. See sequence, next page.

Prevent pipe corrosion by connecting galvanized and copper pipes with a dielectric union fitting. The pipes above show what will eventually happen if you connect these two metals with a standard union fitting. When connecting copper pipe to galvanized iron pipe, you need to install a *dielectric union* between the two materials. That's because if copper and iron come in contact with each other, a chemical reaction between the metals will increase corrosion. The dielectric union will prevent this corrosion from occurring.

How to make a dielectric union

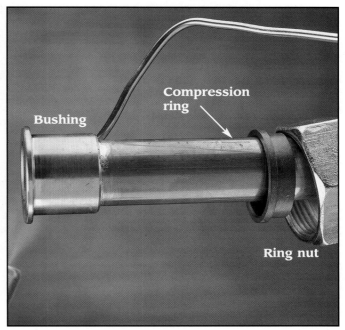

1 Solder the copper or brass bushing from the dielectric union onto the copper pipe using normal soldering techniques: apply flux, insert the fitting, heat the pipe for several seconds until the flux begins to sizzle, then quickly push the solder wire into each joint. The soldered joint should have a thin bead of solder around the lip of the fitting. Make certain not to damage the plastic spacer with the heat from the torch.

2 Screw the galvanized portion of the union onto the galvanized pipe. First apply a bead of pipe joint compound on the threaded end of the pipe. Use two wrenches to apply the fitting: one to hold the galvanized iron pipe steady and the other to turn the union fitting.

3 Connect the parts of the dielectric union. Make certain the spacer is properly aligned, then tighten the ring nut that draws the dielectric union together, using two wrenches.

Hot water nipple

Designed mainly for use with water heaters, the hot water nipple has expandable seals on each threaded end to help it stay watertight throughout dramatic temperature changes. Typically a galvanized nipple with two male-threaded ends, it is best used with a hot water transition fitting (right).

Hot water transition fitting

A threaded adapter lets you transition from a hot water heater nipple to plastic water pipe. Special adapters are needed here since different materials expand at different rates when exposed to heat. The fitting transitions from ¾ in. iron pipe to a ¾ in. CPVC solvent-weld slip.

Banded couplings join pipe to pipe or pipe to tubular trap pieces and tailpieces. To accommodate different thickness pipes, buy a coupling where the rubber or neoprene sleeve beneath the band changes thickness. **Mission couplings,** such as the one shown here, have stickers with abbreviations for plastic (PL) and cast iron (CI) after the listed nominal diameter. Plastic trap adapters and Schedule 40 plastic traps go by the PL designation, as does galvanized steel, which has the same outside diameters as plastic DWV. "Tube" means the fitting is for tubular brass and plastic, which includes tubular PVC, ABS and chromed brass trap arms and tub tailpieces. Sometimes this will be described as a "bathwaste tailpiece" size.

Brass Nipple

The brass nipple is not technically a fitting, but it is invaluable for transitioning from a water heater or a drop-eared 90 in the wall to an angle stop or tub spout. To save money, some plumbers will use galvanized steel nipples instead of brass, which leads to galvanic corrosion and clogged valves.

Trap adapters

Trap adapters are a kind of compression fitting that secures tubular pipe inside standard pipe. Tubular pipes are measured by their outside diameters, so they can slip inside DWV pipes, which are measured loosely by their inside diameters. Accessible traps, like those under sinks, are made of tubular brass or plastic and are held in the drainpipe stub with a trap adapter. The trap adapter may have female threads and be screwed to an iron drainpipe stub, or it may have a female slip cup and be solvent welded to a plastic stub out. Trap adapters can also be attached to most DWV pipe ends with a mission coupling.

Soldered copper and Schedule-40 plastic traps, such as those found in concealed spaces under tubs and showers, often have standard, not tubular, DWV dimensions. In such cases, the trap adapter is used to secure the tubular tub tailpiece into the trap inlet. If the trap is accessible, through a panel or under a sink, a Schedule 40 trap may still have a union nut for cleanout purposes just like a slip nut tubular trap. The trap arm is held to the galvanized waste line with a mission coupling. See pages 96 to 101 for more information on trap adapters.

Trap adapter

Valves

Valves are plumbing fittings that can be adjusted to control the flow of water. Faucets are considered valves, as are shut-offs and essentially any other fitting with a handle or lever that can be opened and closed.

Following are descriptions and repair suggestions for the most common types of valves for home plumbing (excluding kitchen and bathroom faucets, which are discussed in their own chapters). Turn to pages 20 to 21 for exploded-view photographs of each valve type.

Gate valve. If the handle leaks, you may replace the packing washer or packing string. Use heatproof grease on neoprene parts. Leave valve in fully open position. If water gets past the valve when valve is shut or if the valve fails to open fully, you will need to replace the valve itself. Consider replacing a gate valve on a water supply line with a more reliable full-port ball valve. When replacing valves, follow steps for cutting and joining pipe and fittings of your water-line material (rigid plastic, copper, flexible plastic, or galvanized steel). Use the appropriate transition fittings.

Stop-and-waste (globe) valve. Replace the packing washer or packing string for handle leaks and the stem washer and stem-washer screw if your valve does not completely shut off the water. Coat neoprene parts with heatproof grease. Leave valve in fully open position.

Stop valve for fixtures and appliances. Replace the stem washer if this valve does not completely shut off water flow when shut. This washer may snap on and off without use of a stem washer screw. Replace packing washer or packing string if water leaks from the handle when open. Remember to coat neoprene parts with heatproof

PRESSURE REDUCING VALVE

A pressure-reducing valve is installed after your main shutoff valve to reduce chattering and other negative effects of having the water pressure too high in your supply lines.

grease. Leave valve in fully open position. If you need to replace this valve, see pages 46 to 48.

Saddle valve. These are sometimes used to connect icemakers or water filters to a copper supply line. They tap the line with a spike and seal the puncture with a gasket. Leaks may develop at the handle indicating an O-ring failure or at the joint of the valve and pipe indicating a gasket failure. Coat replacement neoprene parts with heatproof grease before installing. Note: Saddle valves are notorious for leaking between the valve and pipe when the gasket fails. They are not permitted by code in many communities. You may choose to replace a saddle valve with a compression tee off a stop valve for the cold water tap of the sink. Repair the hole left by the saddle valve as described on page 30.

Hose bibs (exterior faucets, also called "sillcocks") and some shutoff valves use a compression-style stem washer to block water flow

through the pipe or spout and a packing washer to keep water from flowing out along the spindle under the handle. Fix these valves like you would other compression faucets. If you can't fix a shut off valve, replace it with a ball valve.

Pressure-reducing valves (See photo, above). If your water pipes are excessively noisy, your water pressure may be too high. High water pressure from a municipal supplier can cause chattering and pounding of pipes. Excessively high water pressure can also produce leaks, drippy faucets, and wear out solenoid valves in washing machines and dishwashers. Pressure gauges are available from plumbing supply stores and sellers of sprinkler systems. Purchase or borrow one and thread it onto an outside hose bib to see if your water pressure is more than 80 PSI. You may install a high quality pressure-reducing valve after the water meter—one made from non-ferrous parts or non-ferrous and stainless steel parts.

COMMON VALVE TYPES

STOP & WASTE VALVE
(globe valve)

GATE VALVE

BALL VALVE

 Handle nut

 Handle

 Packing nut

Handle nut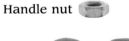

Packing washer

 Handle

 Packing nut

Packing nut

 Packing nut

 Spindle

 Packing washer

Gasket

Stem washer

Spindle

 Spindle

Ball

Stem screw

 Gate

Valve union

Valve body

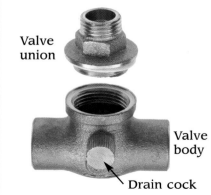

Valve body

Drain cock

Valve body

Valve body

Handle

COMMON VALVE TYPES

HOSE BIB
(SILLCOCK)

SHUT-OFF VALVE

SADDLE VALVE

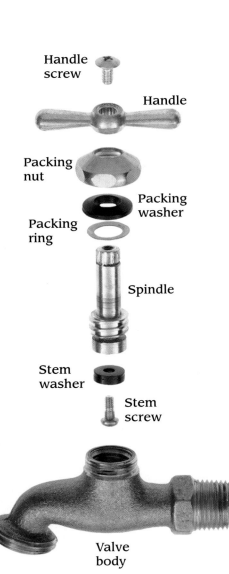

Handle
screw

Handle

Packing
nut

Packing
washer

Packing
ring

Spindle

Stem
washer

Stem
screw

Valve
body

Handle
screw

Handle

Packing
nut

Packing
washer

Spindle

Stem
washer

Valve
body

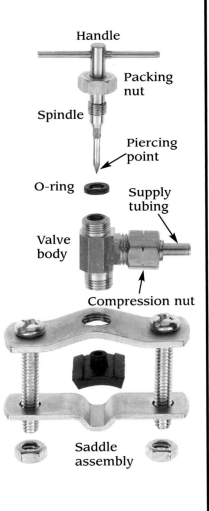

Handle

Packing
nut

Spindle

Piercing
point

O-ring

Supply
tubing

Valve
body

Compression nut

Saddle
assembly

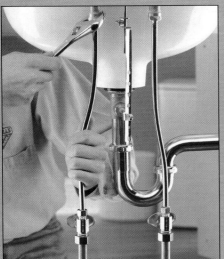

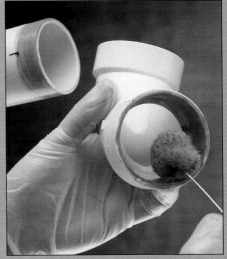

Plumbing Skills

Much of the success or failure of any plumbing project comes down to basic plumbing skills. If joints are fashioned incorrectly or poorly, they will leak. If repairs aren't done carefully, they will be ineffective. If pipes are not cut squarely and smoothly, they won't make a good seal.

This section of *Home Plumbing* covers the basic materials handling information you'll need to know when attempting plumbing projects or repairs. All of the most common plumbing materials discussed in the previous chapter are put into play, from soldering copper fittings, or "sweating" as it's often called (pages 26 to 27) to thawing out and repairing frozen pipes (pages 58 to 59). You'll see explanations on how to replace a section of a cast iron drain stack (page 43), how to

thread galvanized pipe (page 42), and many other skills you may need to work on the plumbing system in your house. And you'll also find sections detailing the best ways to silence noisy pipes (page 56 to 57) or run all new pipes to a fixture (pages 54 to 55).

Before you begin any plumbing project, it's always a good idea to practice the skills you'll need. Buy a couple of copper repair couplings (they're very inexpensive) and solder them to a cutoff piece of copper tubing. Experiment with the settings on your propane torch until you make it produce the perfect flame (pages 26 to 27). Once you're confident (and competent) in your basic plumbing skills, your chances of working successfully will increase dramatically.

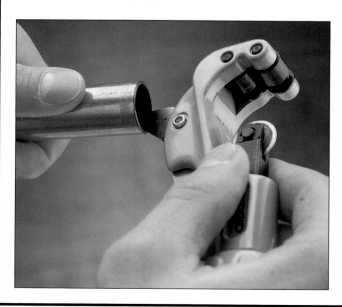

How to cut copper tubing

Copper pipe & tubing

Copper supply pipe and tubing is lightweight, easy to cut, corrosion-resistant and the joints formed by soldering two pieces of copper are extremely strong and durable when made correctly.

There are two basic types of copper tubing: *rigid* and *flexible*. Rigid pipe is used for most of the water supply system. It comes in three grades, ranked according to the thickness of the pipe walls. *Type M* is the thinnest and cheapest copper pipe. It's suitable for most household plumbing uses. *Type L* is generally used for commercial applications, and *Type K* is the thickest rigid copper and is not used in residential plumbing. Flexible copper tubing can be used for water supply lines in some cases, but is used most commonly for gas lines.

Soldering copper fittings. If applied correctly, solder makes a strong and enduring joint between copper pipes and fittings. The most important keys to soldering success are proper preparations of the surfaces and heating the pipes and fittings to a hot enough (but not too hot) temperature with a propane torch.

Before soldering, polish all surfaces to be joined to remove oxidized copper. You may use emery or plumber's sanding cloth, or buy a brush tool that cleans the inside of the fitting and the outside of the pipe. Using the wrong kind of abrasive can remove too much copper or leave shavings that can interfere with the flow of molten solder.

If you are working on an installed supply line, shut off the water and open the valve drain cock, flush the toilets, and empty lines by opening low and high faucets including the outside sillcock. If you shut off and drain the hot water system, make sure you also turn off the water heater. Mark

1 Mark the length you want to cut with a black magic marker. Then place the tubing cutter over the tubing and tighten the handle. The cutting wheel should be aligned with the cutting line and both rollers on the tubing cutter should be making contact with the pipe. Do not overtighten. Turn the tubing cutter one complete rotation, scoring a continuous straight line around the pipe. Rotate the cutter in opposite directions, tightening the handle slightly after every two rotations (inset photo) until the cut is complete and the waste piece of tubing drops free.

2 Remove burrs from the inside lip of the tubing with the reaming blade found on the back of the tubing cutter, or use a rat-tail file. Removing all the burrs and creating a perfectly smooth edge is critical to successful soldering. It also ensures good water flow and fewer deposits over time.

where your new or old tubing sections are to be cut with a small hacksaw. Cut the ends off new lengths of tubing, which may be misshapen. Measure new tubing to fully insert in the fitting sockets, but not so long that the pipe will be compressed or cocked.

Tips for soldering:

- Wear latex or rubber gloves and protective eyewear when soldering.
- Never leave a cup on a fitting empty while soldering, as it will tin over with solder.
- To avoid getting melted solder in nearby joints, wrap the joints with wet rags before sweating the new connection.
- When heating the copper joint, do not tip the propane fuel container of your torch upside down.
- One indication that metal is hot enough to melt the solder is the formation of sweat beads near the pipe joint.
- For ½-in. tubing, use about a ¾ in. length of solder from the coil per joint. For ¾ in. tubing, use about a 1⅛ in. strip. The correct amount of solder is 1½ times the diameter of the pipe.
- If the solder melts and forms beads instead of moving into the joint, or if the flux burns off, your joint is too hot. You may have to start again with new materials. Copper that has been overheated is difficult to solder.
- Let the solder solidify to a frosty color before touching the pipe so you don't crack the joint.
- Practice soldering and other techniques on scrap before performing them on your plumbing.
- Do assembly work outside of the wall on a workbench and then install the assembly with repair couplings.
- Always keep a fire extinguisher close at hand when soldering copper materials.

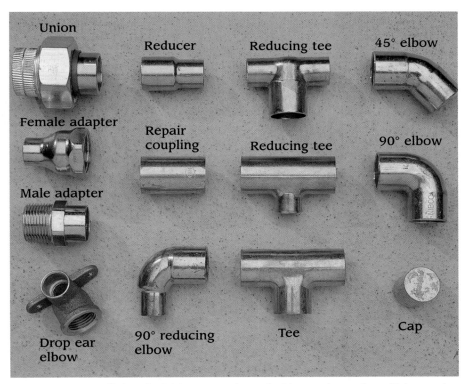

The most common fittings for joining copper pipe and tubing are shown above. Make sure the inside diameter of the fitting matches the outside diameter of your tubing.

How to prepare copper fittings & tubing for solder

1 Clean the ends. Copper pipes, tubing and fittings must be clean and burr-free before soldering. A brush tool allows you to clean both the inside and outside surfaces, and to scuff the copper slightly, creating a bonding surface for the solder.

2 Clean the fitting. Here, a brush tool is used to clean the inside of an elbow fitting. Avoid touching the mating areas of the pipes and fittings—the oil residue from your skin can interfere with the bond. For best protection, wear rubber gloves.

How to solder copper tubing

1 After tubing is clean and dry (See previous page), apply a thin layer of soldering flux (sometimes called soldering paste) to the ends of the tube. The flux cleans and prepares the copper surface for the solder. It should cover about 1 in. of the tube end. Stir the flux before applying.

2 Apply flux to the inside, mating surface of the fitting, using the flux brush. Avoid touching surfaces in the joint area.

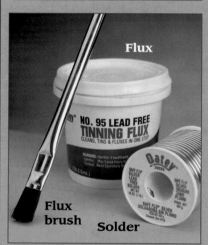

Flux

Flux brush   **Solder**

Propane torches & supplies

An auto-igniting propane torch can help you solder more quickly and efficiently. The torch produces a flame with the push of a button, saving fuel and allowing you to work faster. A propane torch with an auto-igniting starter is also safer than a manually-lighted torch. There is less chance of a flare-up when starting and the flame is eliminated when you release the button. To solder, you'll need flux, solder wire and a brush to apply the flux.

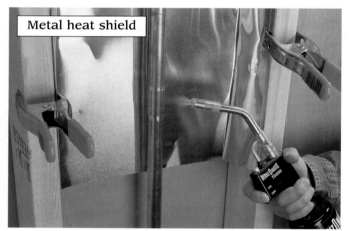

Metal heat shield

"Torch guard" cloth

Heat shields

A propane torch can easily ignite nearby flammable materials when soldering. If you are working on pipes near anything flammable, use a heat block. An effective block can be fashioned out of sheet metal (at least 26-gauge). Clamp the metal to nearby studs. Be warned, however, that the metal can become very hot. Another option is to use specially designed "torch guard" cloth behind pipes you are soldering.

TIP: During repairs, protect surfaces to be soldered from water drips by stuffing white bread into the openings of wet pipe. Do this just before soldering, since the bread will only hold up for a little while. Never use whole wheat bread or bread with grains or seeds in it. White bread will disintegrate when the water supply is turned back on.

3 After assembling the joint so it fits together snugly, use a propane or MAPP torch to heat the fitting—not the tubing—for several seconds. The torch flame should be mostly blue in color and 1 to 2 in. long. The flux should begin to sizzle when the proper temperature is achieved. Heat all sides of the fitting.

4 Before heating the joint, prepare the wire solder by unwinding about 1 ft. of wire and bending the first 2 in. into a right angle. Then, when the joint is hot, touch the solder to the tube. If the solder melts, the tube is ready. Remove the torch and quickly push about 1 in. of solder wire into the joint. You don't need to move the solder around the tube—capillary action will draw the melted solder into the joint. A thin bead of solder should form all the way around the lip.

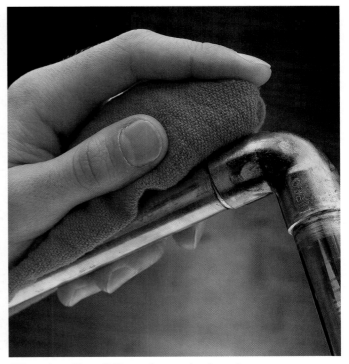

5 Let the solder solidify to a frosty color, then wipe away any excess solder from the joint with a dry rag. Take care: the copper will be extremely hot. When the tube has cooled, turn on the water and check for leaks. If water does seep from a joint, drain the lines. Resist the urge to reapply additional soldering paste to the rim of the joint and reapply solder. This fix usually fails. The best chance for success is to remove the fitting, clean the tubes and try again with a new fitting.

How to splice a tee fitting to an accessible pipe

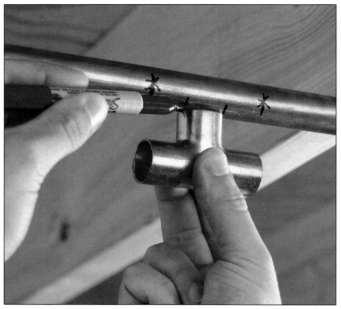

1 Determine where the tee needs to be inserted on the supply line, then mark the full length of the tee onto the run of the pipe. Mark the outside diameter of the tee outlet. Cross out the outer marks to avoid potential confusion. Cut the pipe with a tubing cutter at these inner marks and ream and polish the cut ends. Remember to also polish the inside of the tee.

2 Apply flux to the fittings and pipe ends. Pull the legs of the pipe out of alignment to slip the tee on one leg. Flex and lift the pipes out and apart to slip the other leg into the tee. Install the branch line or stub in the tee outlet after preparing it for solder. Solder all of the joints according to the instructions on pages 26 to 27. NOTE: On a vertical run, it is easier to start with the top joint and work down. The bottom joint will take less solder since solder will migrate down from the other joints.

How to splice a tee fitting to a hard-to-reach pipe

NOTE: If the method shown at the top of this page will put too much strain on your pipes, or if the spot where the tee needs to be inserted is against a joist or hard to reach, your best bet is to make an assembly of the tee and three pipe stubs, then solder the assembly into the existing line using repair couplings. Repair couplings have no stops, so they may be conveniently slipped out of the way onto a pipe leg while you are positioning the assembly.

1 Solder three pipe stubs into the tee fitting on your bench. Then, position the assembly in the correct position in the work area and mark the run of your assembly onto the pipes that will be cut.

2 Cut the pipes at the marks and slip repair couplings over each free end, sliding them back and out of the way. Insert the tee assembly between the free pipe ends and slide the couplings over the joints so they are centered over the joint. Mark the position on the pipes. Make sure a vertical stub is aligned in a vertical position, then solder the assembly and couplings to the line pipes.

Soldering brass valves

Valves have rubber, neoprene, and Teflon parts that can be damaged by heat. On some valves, you can remove non-metallic parts before soldering. Otherwise, you should solder one side as quickly as possible, let the valve cool for fifteen minutes, and then solder the other side. With large valves, such as tub and shower valves, use threaded transition fittings or unions to avoid having to heat the large metal casing of the valve at all.

If the valve you're planning to solder into your copper line can be disassembled, do it before soldering, then reassemble the valve after the body is securely in place and the fitting has cooled. In most cases, removing the retaining nut that secures the valve stem is the key to disassembly.

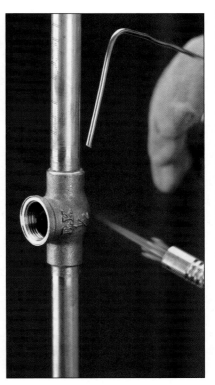

Heat the brass valve body, not the adjoining copper tube. Brass requires longer exposure to the torch flame to attain sufficient heat to melt and draw in the solder.

HOW TO DISASSEMBLE A SOLDERED JOINT

1 If a soldered joint fails, the best solution is to break the joint, clean up the pipes, and try again with a new fitting. First, turn off the water supply in the house and drain the pipes by opening the highest and lowest faucets. Then, heat the fitting using a propane torch. Hold the flame to the fitting until the solder begins to melt.

2 Using slide-jaw style pliers, grip the pipe and pull it free from the fitting before the solder resets. Be careful to avoid any steam trapped inside the pipe.

3 Allow the pipe to cool (you can speed cooling by wrapping the pipe in a damp rag). Clean the old solder residue from the pipes using emery paper and a pipe brush. Do not reuse fittings.

Repairing leaks in copper tubing

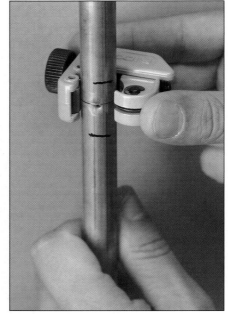

1 If the hole is small, polish a band the full circumference of the pipe two or three inches on either side of the hole, using emery cloth. Center a repair coupling in front of the pipe on the hole and mark the pipe where the edges of the coupling fall. Use a longer coupling for a large hole or tear or use a stub of pipe and two short couplings.

2 Cut the pipe squarely at the hole for a small hole. If you are using a longer repair coupling and cutting out a damaged section, make sure to cut the pipe to leave at least ¼ to ½ in. of overlap between the pipe and the coupling (or couplings) at each end.

3 Try to pull the pipes off center enough to slide the coupling onto one free pipe end. If you can't do this, you will need to replace a short section of pipe, inserting the replacement with two repair couplings (See page 28, bottom). Remove the coupling and prepare it and the pipe legs for soldering.

4 Slide the coupling over both legs of pipe, centering it between the marks made on the pipe. Solder both legs into the coupling at once by distributing heat over the length of the fitting with your torch.

If your pipe has a small, clean hole not located too close to a fitting, and the pipe is not dented, you may install a single, small repair coupling to cover the damaged area. With larger damaged areas, you'll need to remove a section of pipe beyond the damaged area with a tubing cutter or hacksaw and replace the section with either a long repair coupling or a section of pipe and two repair couplings. It is important to use repair couplings, not stop couplings (See page 15), since you will need to slide the couplings entirely on to one of the pipe ends being joined in order to align the sections of pipe.

Before starting a repair, locate and determine the extent of the damage to the leaking pipe. Turn off the water (or the valve to the hot water) and drain down the hot or cold water by opening all spigots, drain cocks on valves, and the exterior sill cock for cold. Flush the toilets for cold and turn off the water heater for hot. By opening up your water system, water can flow out through low valves while air enters through high valves.

If the repair will interfere with an existing fitting, it is necessary to replace that fitting with a new one of the same kind, even if the fitting itself is not damaged. Your joints stand a greater chance of success with a fresh fitting.

Emergency repairs. Your copper pipe just froze and burst, or perhaps it was punctured. If the damage does not involve a fitting, you may apply a temporary fix with a sleeve clamp. After stopping the leak, you will need to prevent the pipe from freezing again (See pages 58 to 59). Repair the pipe permanently as soon as you can.

How to replace a damaged fitting

1 Cut out the damaged area by cutting all pipes entering the fitting near the leak. Make sure to cut the pipe squarely and in places that are convenient to make the repair.

2 Using your torch, a vise, and slide-jaw pliers, detach tubing pieces from the old fitting (See page 29). Use them to determine the lengths of replacement tubing. Cut new lengths of tubing and prepare these and a new fitting for soldering. Do not reuse an old fitting. Solder the tubes into the fitting to match the assembly you removed.

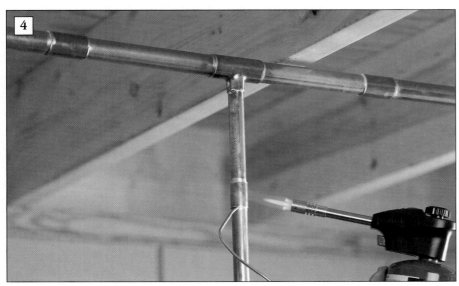

Replacement assembly

3 Determine if your repair assembly will fit. If it won't, modify or rebuild it. Score marks one half the length of your repair couplings onto existing pipe legs and the tubing ends of your assembly.

4 Prepare all tubing ends and couplings for soldering. Slide the couplings onto the tubing ends of your assembly. Join the assembly to existing pipes, aligning the couplings between your marks. Solder all joints according to the instructions on pages 26 to 27.

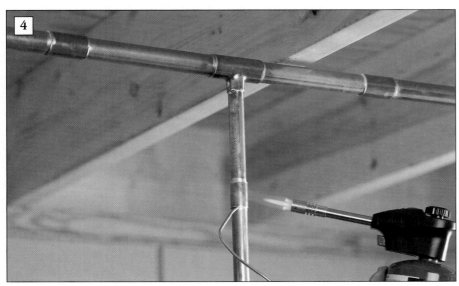

HOW TO MAKE AN EMERGENCY REPAIR WITH A PATCH KIT

1 Turn off the water at the main shut-off valve and open faucets above and below the leak. Use a file to smooth torn areas of metal pipe. If the pipe is frozen, use a hair dryer or a heat gun on low to thaw it (See pages 58 to 59). Once the water flow is under control, head to the hardware store and purchase a pipe patching kit (See below). Make sure you know the diameter of your pipe.

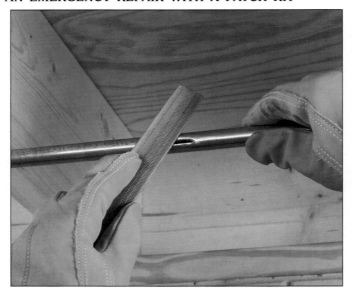

2 The specific installation instructions may vary by kit manufacturer. For the kit shown here (a common one) a piece of neoprene rubber is positioned over the damaged area then clamped to the pipe to seal the leak temporarily. Center the rubber patch over the leak with the seam pointed away from the leak.

3 Secure the clamping device over the patch, usually by tightening screws. Remove the patch and make a permanent repair (See previous pages) as soon as possible.

TIP: If you are truly in an emergency situation or are unable to locate a packaged pipe patching kit, a workable, short-term alternative is to wrap the pipe with a piece of rubber or even reinforced plastic. Tighten several hose clamps over the patch material.

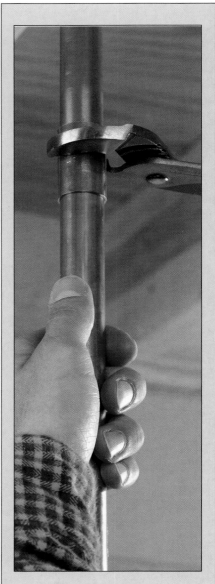

Crimp a fitting to keep it from sliding

If it is necessary to hold a pipe in a fitting before you solder the parts together, squeeze the fitting lightly but firmly with pliers to crimp it onto the pipe. Then, twist the coupling slightly, taking care not to alter its correct alignment with the pipe. The slightly elliptical shape of the fitting resulting from the crimp will create just enough of a tension fit to hold it in place, without affecting the quality of the connection.

Flare fittings

Flare fittings work similarly to compression fittings, except that instead of a brass ferrule, the flared end of the pipe itself is compressed between the fitting and the nut. Flare fittings are not used much with water supply pipes and tubes anymore, but they are still a common method for connecting flexible copper gas and fuel lines. When attaching a flare fitting on flexible gas tubing, tighten it one rotation past finger tight. Use one wrench to keep the fitting body from rotating.

Test a gas fitting for leaks by applying soap bubbles when the gas is back on. Tighten a bubbling joint until the bubbling ceases.

If you need to replace a flare joint with compression joint on a water distribution pipe below a fixture, cut off the flared pipe end.

Flare nut

Brass flare union

Flare nut

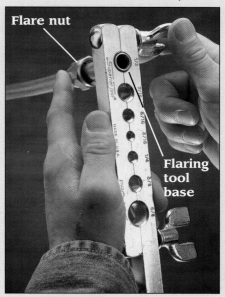

Flare nut

Flaring tool base

1 Flare fittings are most often used for flexible copper gas lines. When flare fittings are used on water supply lines, the connections must not be concealed by walls. To flare copper, you'll need a flaring kit and brass flare fittings. Slide the flare nut over the tubing, then clamp the tubing into the flaring tool. The tubing end must be flush with the face of the flaring tool base. NOTE: Do not attempt to use a flare fitting with rigid copper—flare fittings should be used with flexible copper tubing only.

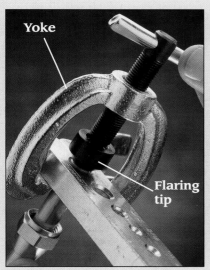

Yoke

Flaring tip

2 Slip the yoke part of the flaring tool around the base. Center the flaring tip of the yoke over the pipe. Turn the handle until it stops against the face of the flaring tool base, resulting in a flared end on the tubing. Oiling the flaring tip will produce a smoother flare.

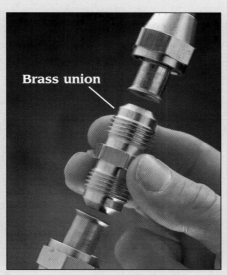

Brass union

3 Flare the end of the other tube, then place the brass union between the flared ends.

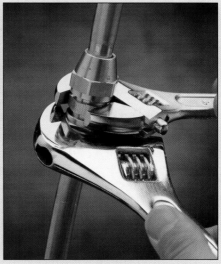

4 Tighten the flare nuts onto the union using two pliers or wrenches and turning in opposite directions. Joint compound is not necessary. Check for leaks. If the fitting leaks, tighten the nuts slightly.

Plastic pipe

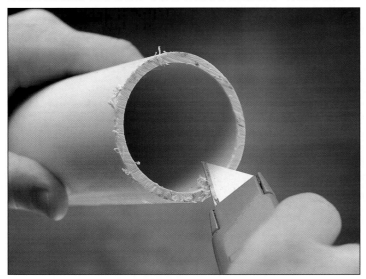

After cutting the pipe (See next page), trim away burrs on the cut edges. Plastic pipe must be smooth and free of burrs to ensure a watertight connection. Use a utility knife to slice off burrs from the edges of the cuts.

Scuff and degloss mating surfaces of pipes. It's a good idea to lightly sand the outside of the pipe and the inside of the connection hub using sandcloth or emery paper before applying primer. Surfaces that are to be glued together should have a dull finish.

PVC and other types of plastic pipe are more popular today than ever before. As a building material, PVC pipe is inexpensive, easy to work, lightweight and relatively forgiving. PVC and ABS are rigid forms of plastic pipe that are used almost exclusively for drain and vent systems. CPVC is a rigid plastic that's used for water supply systems—it resists heat better than ordinary PVC. Other types of plastic pipe include PE, used mostly for exterior, underground installations, and PB, a flexible plastic that's also used for supply pipe but has experienced a decline in use.

Except for flimsy plastic drain trap parts, which are joined with slip nuts and nylon washers, plastic pipe is connected by solvent welding. Solvent welding is a lot like welding metal. Your goal is to liquefy the pipe and fitting surfaces then let them harden together. Primer makes the surface of the pipe and the inside surface of the fitting soft or liquid. Cement helps keep the pipe material liquid, but it also has PVC, CPVC, or ABS suspended in it, which fills in any gaps. Naturally, the cement has to be applied when the fitting and pipe are still damp with primer, so that the plastic on the joining sur-

Materials for solvent-welding PVC

When working with PVC pipe, make certain you use the correct materials. Solvent and primer should be labeled for PVC pipe (ABS requires ABS cement and no primer). Colored solvent and primer allow you to inspect more easily to make sure all parts are coated. Sandcloth or emery paper is used to smooth rough edges. Primer helps degloss the pipe's slick surface, ensuring a good seal. If you buy more than a quart of cement, transfer some to a quart container for application, since the cement is extremely volatile. Not only will a gallon become useless before you finish it, but you will be breathing more fumes. Plastic pipe cements are toxic. You should

wear a long sleeved shirt, vinyl or latex gloves, eye protection, and use air circulating fans or a respirator. Keep the cover on the cement when you are not using it.

faces is receptive to the cement. With ABS, which is more reactive, the solvent in the cement is enough to dissolve the plastic and no primer is used.

Joining rigid plastic drain pipes can be difficult. Expect that you'll make some mistakes at first and have to cut out and redo some joints. ABS drainpipe tends to set quickly. It's a good idea to have a friend to help twist the pipes into place. PVC set time is extremely temperature dependent. In cool weather you may have to hold the pipe three minutes for the larger drain pipes to set. In hot weather you will barely have time to get the joint in position before it sets up. Keep PVC in the shade in hot weather. Before using plastic pipe, bring it close to room temperature. Protect plastic pipe from extended exposure to direct sunlight. Eventually, the UV radiation can weaken the material.

PVC PIPE FITTINGS

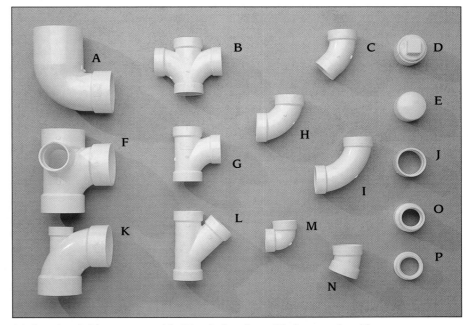

(A) closet bend; (B) waste cross; (C) 45° reducing elbow; (D) cleanout plug; (E) cap; (F) waste tee with side inlet; (G) waste/sanitary tee; (H) 90° elbow; (I) long-sweep 90° elbow; (J) coupling; (K) 90° elbow with side inlet; (L) wye fitting; (M) vent elbow; (N) 22° elbow; (O) reducer; (P) reducing bushing.

METHODS FOR CUTTING PVC PIPE

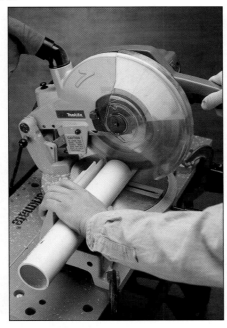

Plastic tubing cutters

A PVC ratchet cutter will make short work of smaller PVC pipes. This tool is especially useful if you have a large variety of plastic pipes to cut.

Power miter saw

If you have a lot of plastic pipe to cut—and you want to make neat cuts very quickly—use a power miter saw fitted with a fine blade that has a high number of teeth per inch.

PVC pipe saw

To cut PVC pipe by hand, a PVC pipe saw will give you better results than an ordinary hacksaw. Make certain to hold the pipe securely in a vise and keep the saw blade straight while cutting.

1 Cut the pipes to length and prepare the mating surfaces (see previous pages). Fit the pipes and fittings together in the desired layout. Draw an alignment mark across each joint with a permanent marker.

2 Apply PVC primer to the outside of the pipe and the inside of the connection hub or fitting. The primer is colored so you can see when full coverage has been achieved. Wear disposable gloves, and make sure the work area is adequately ventilated. Also be sure to read the directions and safety precautions on the labels of all products you'll be using.

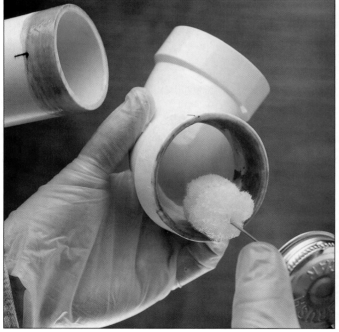

3 Apply a thick coat of solvent glue to the outside of the pipe and a thin coat on the inside of the connection hub.

4 Quickly slip the pipes and fittings together so the alignment mark you drew across the joint is about 2 in. off center, then twist the pipes into alignment. This will ensure the solvent is spread evenly. The solvent will set in about 30 seconds, so don't waste time. Hold the pipes steady for another 20 seconds (longer in colder weather), then wipe away excess solvent with a rag. Don't disturb the joint for 30 minutes.

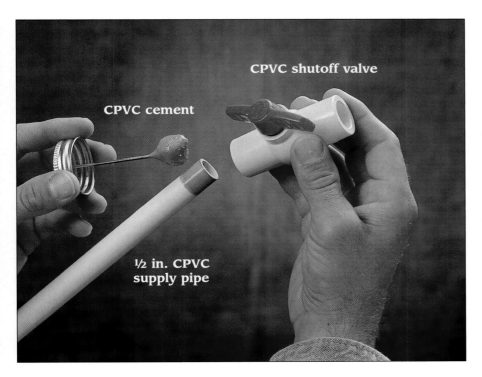

CPVC cement

CPVC shutoff valve

½ in. CPVC supply pipe

CPVC supply pipe growing in code acceptance

Until recently, most water supply pipe used to hook up plumbing fixtures has been copper or stainless steel. CPVC (chlorinated polyvinyl chloride) supply pipe has been around for some time, but many local codemakers have been reluctant to accept it. That's changing. CPVC has been found to be a reliable supply pipe material that's easy to work and also has better insulating power than other materials. It's installed using the same solvent welding process as PVC (See previous page), but requires CPVC primer and cement. Check with your local building inspection department to find out if CPVC supply pipe is allowed in your area.

Code tips for working with plastic pipe

- Hose bibs and nipples for tub spouts and other fixtures need to be connected to metal components anchored to the building. Transition from CPVC prior to the anchored fitting. Stub outs may be CPVC.

- Plastic pipe should not be exposed to direct sunlight in its installed location. The exception is plastic vent chimneys, which may be protected with latex paint.

- There should be at least 6 in. of metallic piping between a water heater and plastic pipe. CPVC cannot be placed downstream of an instantaneous water heater (coil or immersion).

- Don't solder closer than 18 in. to an installed plastic-to-metal adapter on a copper water line.

- Don't tighten a male metal fitting into a plastic fitting. When joining plastic to plastic with a threaded fitting, be aware that a small diameter pipe may be over tightened by hand alone. The female side of overtightened plastic pipes may crack immediately or later. Larger DWV pipes are tougher. They may require up to a full turn past hand tight using a strap wrench. Lubricants, pipe joint compounds, and putties shall not be used unless the label says they may be used with the type of plastic you are using.

ELECTRICAL GROUNDING TIPS FOR PVC PIPE

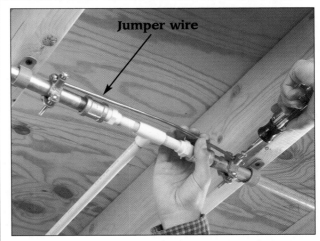

Jumper wire

(ABOVE) If you insert plastic fittings or piping into an existing metal line, bridge the metal pipe ends with a jumper wire in case the pipes are used as an electric ground.

(RIGHT) The permanent sign: "This building has non-metallic interior water piping" must be affixed to the electric service panel.

NOTICE
This building has non-metallic interior water piping.

Galvanized pipe

Galvanized pipe is rarely installed in new construction today for one main reason: it's time-consuming to install. Compared to soldering copper tubing, or solvent-welding plastic pipe, putting together galvanized pipe is slow work. However, replacing a section of old galvanized pipe with new galvanized pipe is a reasonable project for the do-it-yourselfer.

Galvanized pipe is connected with threaded joints. You'll need to rent a pipe vise, a reamer and a threader to thread your own galvanized pipes at home. Get a threader with a head that is the same nominal diameter as the pipe you are planning to thread. You'll also need a bottle of cutting oil.

Like all metal pipe, galvanized iron will eventually corrode and need replacing. But replacing an entire system of galvanized pipe is a big, time-consuming job. Remember that with galvanized iron, you cannot simply unscrew a middle section of piping without first disassembling the entire run.

Occasionally, however, galvanized pipe will corrode in just a small area. How do you replace the damaged section without removing a whole run of piping? Simple. Use a three-piece union. The union will allow you to sidestep the laborious job of disassembling an entire run of pipe.

When shopping for replacement pipe, specify the nominal diameter of the pipe you need. Pre-threaded pipes, called nipples, are available in lengths up to 1 ft. or longer. For longer runs, have the store cut and thread the pipe to your dimensions. You can also thread your own by following the steps outlined on the next page.

One warning: Galvanized iron pipe, which has a silver color, is sometimes confused with "black iron" pipe. Black iron pipe is used only for gas lines. Never use it for water.

Galvanized pipes are susceptible to corrosion (as you can see in the example above) and are more likely to become encrusted with minerals than other pipe types. Mineral precipitates and pipe material that dislodges from pipes can clog appliances and fixtures, especially after repairs are done to the plumbing. You can improve the performance of individual appliances and fixtures by periodically cleaning filter screens at water inlets and the aerators of faucet spouts, and occasionally cleaning out the internal parts of faucet valves and fill valves on toilets. If your house has galvanized water pipe and you experience problems with low water pressure or leaks, it might be time for all new plumbing.

GALVANIZED PIPE FITTINGS

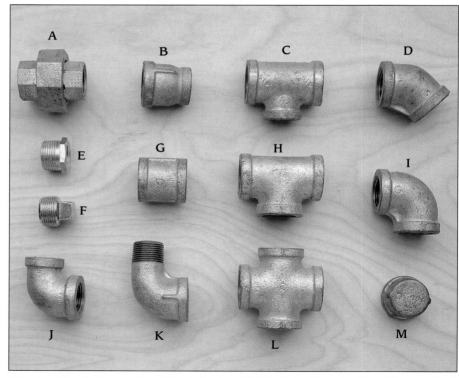

(A) union; (B) reducing coupling; (C) reducing tee; (D) 45° elbow; (E) hex bushing; (F) square head plug; (G) coupling; (H) tee; (I) 90° elbow; (J) 90° reducing elbow; (K) 90° street elbow; (L) cross connector; (M) cap.

HOW TO REPLACE A SECTION OF GALVANIZED PIPE

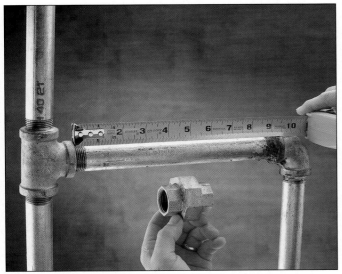

1 Shut off water. Measure the length of replacement pipe you will need. Make certain to include ½ in. for each threaded end that inserts into a fitting. However, subtract the width the union adds from the measurements. When assembled, the union and replacement pipe must equal the length of the section being removed.

2 Cut through the old galvanized pipe with a hacksaw or reciprocating saw fitted with a metal-cutting blade.

3 Remove the corroded pipe with a pipe wrench. If the fitting is stubborn, grasp the fitting with a second pipe wrench. If the joint still won't loosen, heat it with a propane torch for five to ten seconds, making certain not to ignite any nearby materials. Once the fitting is removed, clean the pipe threads with a wire brush.

4 To remove corroded fittings, use two wrenches. Face the jaws in opposite directions and use one wrench to remove the fitting while the other wrench holds the pipe in place.

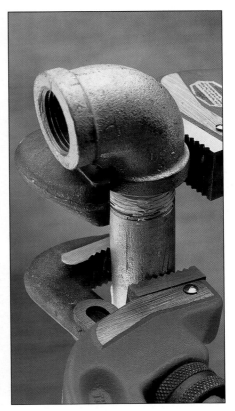

5 To attach a new fitting, apply pipe joint compound on the threaded ends of all pipes and nipples. Screw the new fitting onto the pipe and tighten with two pipe wrenches. Leave the fitting about ⅛-turn out of alignment to permit installation of the union.

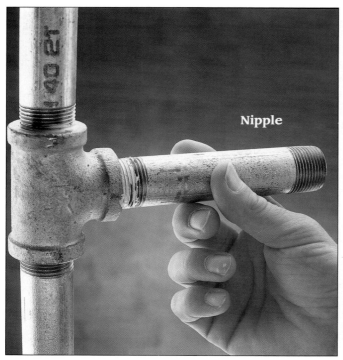

Nipple

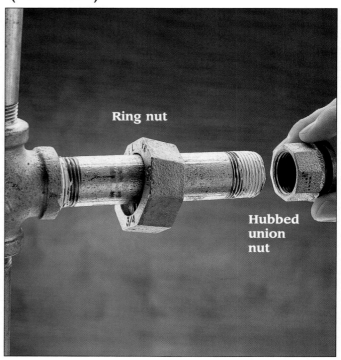

Ring nut

Hubbed union nut

6 Attach a new pipe, or nipple, to the fitting. Apply pipe joint compound or strips of teflon tape to all threads. Tighten with a pipe wrench. Wrap the tape clockwise to keep it from unwinding as you thread the pipe into the fitting.

7 Slide a ring nut onto the installed nipple. Then screw a hubbed union nut onto the nipple. Tighten with a pipe wrench.

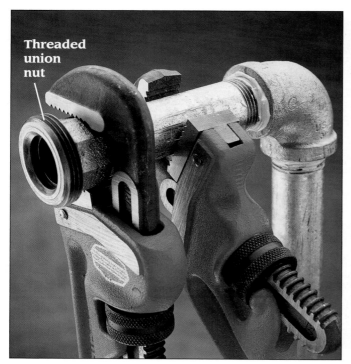

Threaded union nut

Threaded union nut

Hubbed union nut

Ring nut

8 Attach the second nipple to the opposite fitting. Tighten with a pipe wrench. Then screw the threaded union nut onto this nipple. Hold the nipple with a second wrench while attaching the union nut.

9 Align the pipes so the lip of the hubbed union nut fits inside the threaded union nut. Tighten the connection by screwing the ring nut onto the threaded union nut.

EMERGENCY REPAIR FOR GALVANIZED PIPE

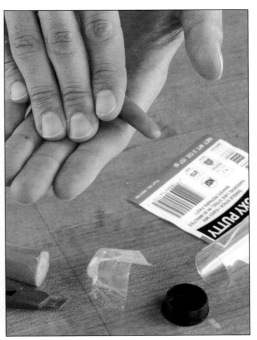

1 If a galvanized steel pipe springs a leak for no apparent reason, it probably means the pipe is rusting out from the inside. Consider replacing one or more sections of pipe. Otherwise, you can temporarily apply an epoxy patch with a pipe clamp. Turn off the water at the main shutoff. Locate the leak. Flush toilets and open faucets and hose bibs to drain the system. Turn off the water heater if the problem is on a hot water line. Use a wire brush and plumbers sandcloth to remove corrosion from the leak area. Wipe clean.

2 Mix the epoxy and hardener to a uniform color according to the product manufacturer's instructions

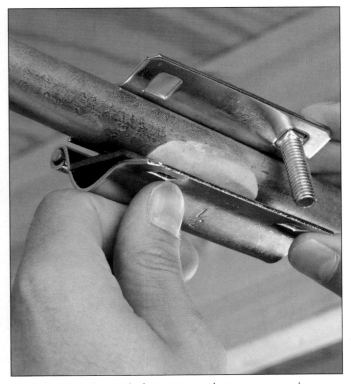

3 If the leak is at a threaded fitting, push the epoxy around the joint of the fitting. If the leak is on the pipe, press the epoxy over the leak.

OPTION: Use a pipe repair clamp to strengthen an epoxy repair on running sections of galvanized pipe (See page 32). Put the rubber gasket and clamp over the epoxy. Make sure the gasket of the clamp presses against the epoxy repair.

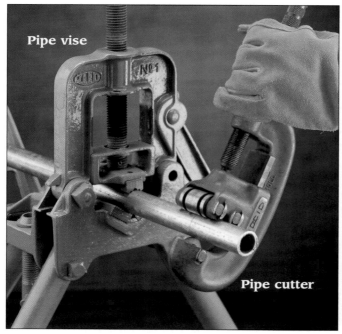

Pipe vise

Pipe cutter

Reamer

1 Before cutting, mark the length you want. Then, secure the pipe in the pipe vise (pipe fitting tools can be rented at most rental centers). Tighten a pipe cutter on the mark and rotate the cutter first one direction, then the other. Tighten the cutter every two rotations until the pipe is cut all the way through.

2 A reamer will remove burrs and jagged edges inside the pipe. Simply insert the nose of the reamer into the pipe, push and turn clockwise. Put a catch basin under the work area to collect pipe shavings and oil as you work.

Threader

3 To thread the pipe, slip the head of the threader over the end of the pipe. Push down on the threader and tighten it until the threader's head, also called the die, bites into the pipe. At that point stop tightening the tool and turn it clockwise around the pipe. Apply lots of cutting oil while turning. Keep turning the threader until the cutting head has cleared the pipe by at least one full turn. If the threader sticks, metal chips are probably blocking progress. In that case, rotate the tool backwards slightly and blow the chips away or remove them with needle nose pliers.

4 When done threading, remove the threader and clean the newly cut threads with a stiff wire brush.

Cast iron pipe

Most older houses, and even some newer ones, have a main stack made out of cast iron. Cutting into the stack to add an auxiliary drain or vent line can be difficult and dangerous. This sequence shows the correct way to support the main stack, cut out a section, and insert a drain or vent connector fitting. Cast iron cutters can be rented at most rental centers.

2 × 4 cleat

Riser clamp

1 Cast iron drain stacks must be braced with a riser clamp at each floor the stack passes through. Make sure the bracing exists, then mark the location for the new connector—the cutout should be sized to the height of the connector, plus 4 to 6 in. of pipe that should be solvent-welded to each end of the connector so it's the same diameter as the stack. Install a riser clamp about 6 in. above the top of the cutout area. Attach cleats to the wall studs to support the riser clamp.

2 Cut out the section of the main stack with a cast iron cutter (sometimes called a pipe breaker). The tool, equipped with cutting wheels on a heavy-linked chain, wraps around the pipe. Ratcheting the cutter wrench handle up and down tightens the chain until the pipe snaps. For safety, it's not a bad idea to support the section to be removed with a riser clamp before cutting it. Make the upper cut first. Remove the waste section carefully.

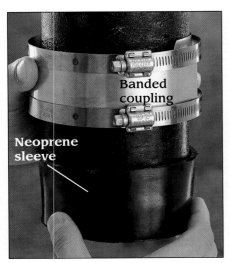

Banded coupling

Neoprene sleeve

PVC connector

3 Slip a banded coupling onto each end of the stack. Slip a neoprene sleeve over the pipe ends. Fold the sleeve out of the way.

4 Insert the connector into the opening, making sure the inlets are oriented exactly as you want them to be.

5 Fold the sleeves over the connectors, making sure the ridge inside each sleeve is flush with the joint. Position the steel bands over the sleeve. Tighten the screw clamps.

Supply pipes & tubes

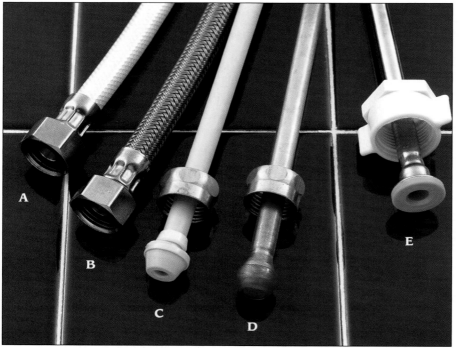

Supply pipes bring water from the water main to the shutoff valves at plumbing fixtures. Regardless of whether they are copper, plastic or galvanized steel, they are either ½ or ¾ in. in diameter. From the shutoff, water is sometimes distributed to the fixture via smaller supply tubes. The type of connection between the supply pipe and the shutoff depends on the material the pipe and fitting are made of. These days, supply tube connections tend to be made with compression fittings.

Because supply pipes are pressurized and connected to a virtually inexhaustible source of water, a leak or burst supply pipe has the potential for causing catastrophic damage. Pay attention to them and be sure you know where the main shutoff is in case of a problem.

Supply tubes carry water from the shutoff valve to the fixture. You may find any of the following types in your home: (A) Vinyl mesh with integral nut; (B) Braided steel with integral nut; (C) PEX (cross-linked polyethylene) with "acorn" and slip nut; (D) Copper tubing with "acorn" and slip nut; (E) Chromed brass with ballcock coupling nut.

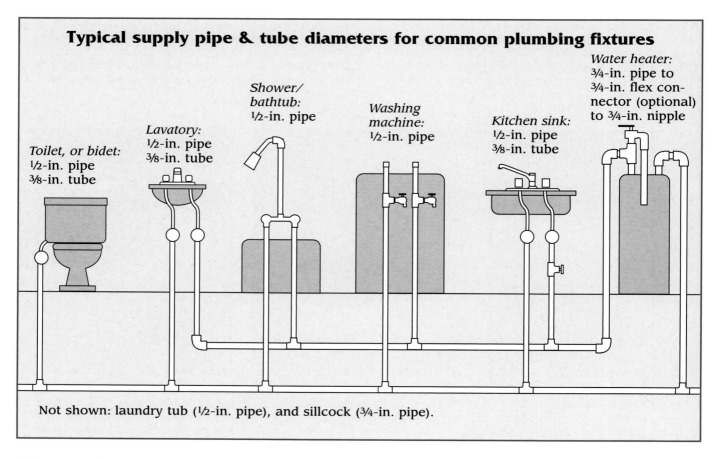

Typical supply pipe & tube diameters for common plumbing fixtures

Water heater: ¾-in. pipe to ¾-in. flex connector (optional) to ¾-in. nipple

Shower/bathtub: ½-in. pipe

Washing machine: ½-in. pipe

Kitchen sink: ½-in. pipe ⅜-in. tube

Lavatory: ½-in. pipe ⅜-in. tube

Toilet, or bidet: ½-in. pipe ⅜-in. tube

Not shown: laundry tub (½-in. pipe), and sillcock (¾-in. pipe).

Repairing leaks between shutoffs & fixtures

Water leaking under the sink (or behind or under any other fixture for that matter) may be caused by a failed supply tube connection or leaky shutoff valve. Fixing a leak here may be as simple as tightening a compression nut or as involved as installing a new supply tube and shutoff. Of course, try the simple fixes first. First, rule out the possibility that the leak is coming from the casting of the sink's faucet, from a leaky faucet or spout, or from the union of a tailpiece and the faucet. If none of these are your problem, look to the supply tubes and stops.

If the leak is at the supply tube joint with the faucet tailpiece, try tightening the coupling nut (See photo, right). You may need to use two adjustable wrenches—one to keep the tailpiece from rotating. Do not use too much force; these nuts and threads can be fragile.

If the valve leaks where it joins the pipe or supply tube, attempt to tighten compression nuts or tighten the valve onto its nipple. Use two wrenches to prevent torquing of the pipe.

If the leak persists, turn off the water supply, unscrew the nut or valve at the leaking joint, and wrap male threads with four layers of Teflon tape in the direction the nut or valve screws on. Wrap the ring with Teflon tape in the direction that the compression nut tightens for compression fittings. Or, apply pipe joint compound to these fittings. If you have a rubber, cone-shaped washer at a connection, replace it. Retighten nuts, turn on water, and see if the leak stops.

If there is a leak-free union at the sink or shutoff valve and the leak persists, replace the supply tube. If it uses a compression fitting, you'll need to get a new compression nut and ring. Bring the old supply tube to the store to get sufficient length and the correct connectors. A braided stainless steel tube can stay long since it's flexible, though you may have to run it in a loop. Compression rings are built in to the captive compression

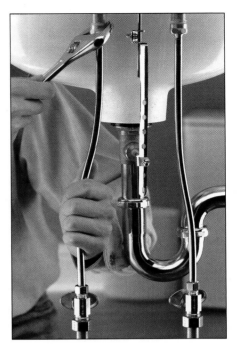

Leaks and drips from supply tubes and shut-off valves are most often caused by a loosened connection. Simply tightening the retainer nuts at the faucet tailpipe or at the shutoff valve is a simple fix that often works.

nut on a braided supply. Acorn head supply tubes should be purchased too long, then cut to fit at the valve end.

The future of water distribution

PEX (cross-linked polyethylene) is a flexible plastic tubing for use with hot and cold potable water. Unlike rigid plumbing systems, PEX is generally run out like wires, with no branching or reducing fittings. Each fixture and appliance has direct hot and/or cold connections to a central manifold. The material is quiet and insulates the water from temperature change. Chemically, it is practically inert, so flavors or chemicals are not imparted to the water and scale will not build up on the insides of the tubes. PEX has been used in Europe for decades and is now gaining popularity in this country. Professionals join PEX with insert fittings, rings, and a special crimping tool.

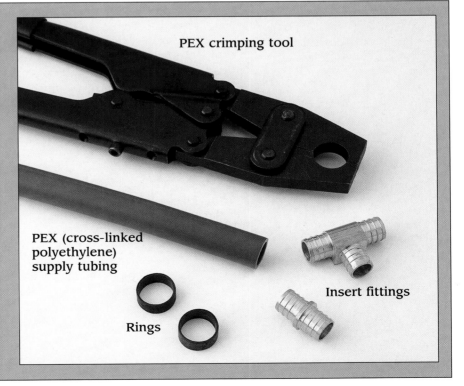

PEX crimping tool

PEX (cross-linked polyethylene) supply tubing

Rings

Insert fittings

Shutoff valves

Shutoff valves can be a little bit tricky since one side connects to a pipe in the wall or floor and the other side connects to a supply tube that goes to the sink or toilet. These are not only different diameter pipes, but may use different kinds of connections as well. When

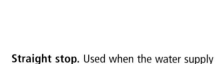

Angle stop. Used when the water supply pipe enters the room through a wall. The 90° angle of the outlet to the inlet allows you to make supply tube connections without kinking the tubes.

Straight stop. Used when the water supply pipe enters the room through the floor. Very common for toilet hookups.

specifying dimensions and joint types, the wall/floor side is called the "inlet" and the supply-tube side is called the "riser outlet." The supply stop itself is called a straight stop or an angle stop, depending on whether it is straight (for a pipe coming out of the floor) or takes a 90° turn (for a pipe coming through a wall).

Here are some terms and concepts you need to know when buying a shutoff valve:

FIP (Female Iron Pipe): The female side of a threaded joint using iron pipe size.

IPS (Iron Pipe Size): A thread standard used with any material, not just iron.

MIP (Male Iron Pipe): The male side of a threaded joint using iron pipe size.

Compression Fitting: A kind of fitting that let's you connect to a smooth pipe without solder. A compression fitting includes a threaded nut and a brass ring called a compression ring that's pressed between the pipe and the nut.

Nipple: This is a short section of IPS threaded pipe.

Nominal (Named) Diameter: All pipes are given a size name, like "½-in. copper pipe." Supply and distribution pipe sizes are named roughly by their inside diameter. This distinguishes pipes from tubes, which are narrower inside than their named tube size.

OD (Outside Diameter): Supply tubes to fixtures and appliances and tubular traps and drain tailpieces are named by their outside diameters (braided supply tubes are an exception).

ID (Inside Diameter): Water pipes, including the nipples and stub outs that come out of the wall and floor, are named roughly by their inside diameters.

Choosing shutoff valves

Following are three scenarios designed to help you choose the correct shutoff valve for your situation.

Scenario #1: *Your hot water comes out of the wall through a threaded brass nipple that measures about ½-in. inside diameter. You want to attach this to a smooth ⅜-in. outside diameter chromed brass supply tube rising to your lavatory sink.*

In this scenario, You need an angle stop with a ½-in. female iron pipe inlet × ⅜-in. outside diameter compression outlet for the tubular riser.

Scenario #2: *Your cold water comes out of the floor through a ½ in. inside-diameter copper pipe. You want to attach this to a ⅜-inch closet (toilet) riser.*

In this scenario, you need a straight stop with a ½-in. compression inlet for copper pipe × a ⅜-in. outside diameter outlet for the tubular compression riser.

Scenario #3: *Your hot water comes out of the wall below the kitchen sink through a ½ in. inside-diameter brass nipple. You need to connect two ⅜-in. supply lines—one to the dishwasher and one to the kitchen sink.*

In this scenario, you can purchase dual outlet supply stops, but these aren't always approved by code. The better solution is to attach a ½-in. female iron pipe tee to the nipple and two close (short) nipples to this, then attach two angle stops to the close nipples. Each stop will have a ½-in. female iron pipe inlet × ⅜-in. tubular compression outlet.

Compression fittings

Compression Fittings. Compression fittings are useful where connections are not permanent. Plumbers use compression fittings mainly to attach small diameter water supply tubes to stop valves beneath sinks, lavatories, toilets and appliances. Some home centers now carry compression fittings for all sizes of water pipes and tubes and in different configurations, such as tees, elbows, and valves. Code usually prohibits concealing compression fittings in walls, since they are more prone to failure than soldered or solvent welded joints. Do not use compression fittings where they will be subject to vibration, which could loosen the compression nut.

NOTE: Compression fittings may be sold by the outside diameter (OD) of pipe. Outside diameter is 1/8 in. greater than nominal diameter of copper and plastic water pipe but is equal to the nominal diameter of flexible copper tubing.

Some compression fittings have compression rings permanently joined to the compression nut. These are preferable since you can't throw the ring out of alignment by overtightening them.

To install a compression fitting on a copper pipe, first cut the pipe with a tubing cutter and ream the inside edge. If you will be joining metal pipe, you may discard the tubular copper inserts (not the compression rings) that come in some compression fittings. Use these inserts with plastic tubing to resist the crushing pressure of the ring.

Next, slide the compression nuts onto the pipe ends, threads facing toward the fitting. Slide the compression rings on after the nuts and coat these with pipe joint compound or Teflon tape. Take care that the pipes enter the fitting straight. Hand tighten the compression nuts, being careful not to cross-thread the nuts and fitting.

Hold the fitting with a wrench and tighten each nut one half turn past hand tight. Do not over tighten or you can cock the ring. After turning the water back on, you may tighten a leaking join just enough to stem the leak.

In addition to making connections between pipes and fittings, compression fittings may be used to join two lengths of pipe. But as with fitting connection, the joint must be accessible. Unless you foresee a need to undo the joint, you're usually better off soldering or solvent welding a permanent fitting to join pipes in a run.

Compression nut

Compression ring

Compression union

Compression ring

Compression nut

How to attach a copper supply tube to a shutoff valve

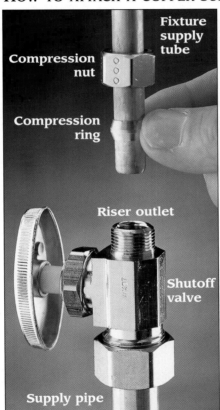

1 With the shutoff valve attached to the supply line (See next page), cut the fixture supply tube to length, allowing 1/2 in. for the portion that will fit inside the shutoff valve. Slip a compression ring and compression nut (usually included with valve) onto the end of the fixture supply tube.

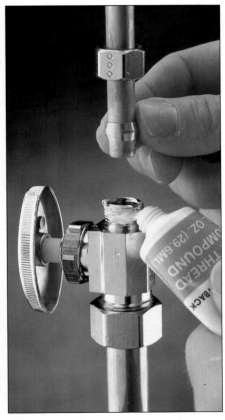

2 Apply pipe joint compound or Teflon tape to the threads of the riser outlet. The compound will serve as a lubricant during compression. Tighten the compression nut 1/2-turn past hand tight over the compression ring at the joint.

HOW TO INSTALL SHUTOFF VALVES WITH COMPRESSION FITTINGS

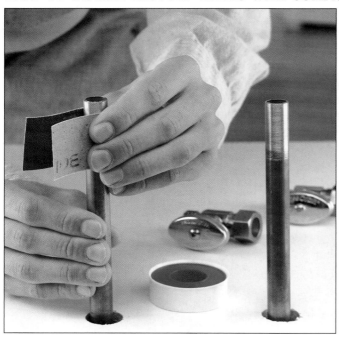

1 Shut off water at the main. Prepare the copper supply pipes by sanding with plumbers sand-cloth or using a pipe brush to remove grime and build-up. Make sure the ends of the pipes are cut squarely.

2 Slip a compression nut, open side up, over a pipe, then fit a compression ring around the tip of the pipe. Fit the compression inlet of the shutoff valve over the compression ring. Apply Teflon tape to the male threads and pipe joint compound to the female threads. Thread the compression nut onto the shutoff valve body and tighten.

HOW TO ATTACH A SHUTOFF VALVE TO A THREADED NIPPLE

1 Shut off water at the main. Make sure the threads on the nipple coming out of the wall are clean, then wrap the threads with Teflon tape.

2 Screw a shutoff valve with female threaded inlet onto the nipple (apply pipe joint compound to the threads first). Hand tighten, then tighten with an adjustable wrench (don't overtighten, though). If using an angled shutoff, make sure the outlet is pointing in the direction of the fixture.

CHROME TUBE INSTALL

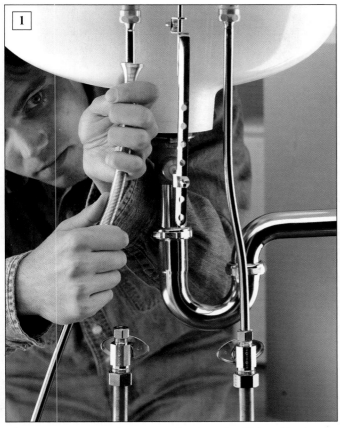

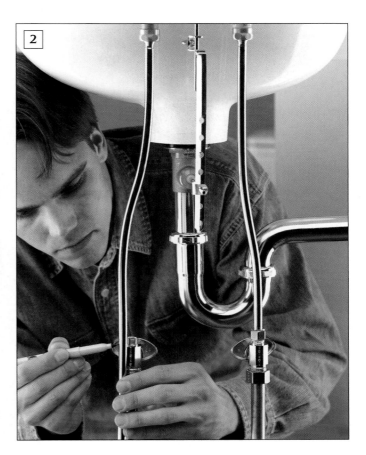

1 Bend uncut chromed copper supply tube with a ⅜-in. tubing bender so that when the acorn end is inserted in the fixture tailpiece, the tube approaches both the tailpiece and the shutoff valve in a straight line. Attach the tube temporarily to the tailpiece while bending, if this is feasible.

2 With the acorn end of the tube secured at the faucet tailpiece, mark the tube with a felt-tip pen where it passes the shutoff valve; include the ½ in. that will be inserted into the valve. Remove the tube and cut it with a tubing cutter by first rotating the cutter one way and then the other as you gradually tighten the cutter head. Gently ream off any rim or burrs left by the cutter.

3 Put the compression nut and compression ring on the pipe, applying pipe-joint compound to the ring. Insert the pipe fully into the valve socket, and slide the ring and compression nut against the threads of the valve. Hand tighten. Tighten the compression nut another ½ to ¾ of a turn, stabilizing the valve if needed with a second wrench. Do not kink the delicate rings by overtightening; you can always tighten a compression nut a little more if the joint leaks when the water is back on. Put pipe joint compound on the coupling nut threads and, inserting the acorn in the tailpiece, tighten the nut onto the tailpiece threads. You may need a basin wrench to accomplish this. Do not overtighten.

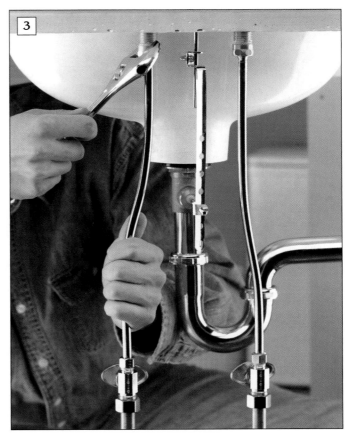

NOTES:
If you are connecting your supply tube to a flexible inlet pipe, you must take special care not to twist the inlet pipe. Hold the fitting of the inlet pipe with one wrench while you tighten the supply nut with another.

Take off the aerator on the spout and leave faucets open until you turn the water back on. When the water is running smoothly, turn off the water and replace the aerator.

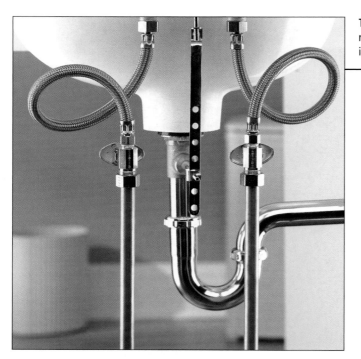

TIP (LEFT): If braided tubes are significantly longer than necessary to reach from the shutoff to the faucet (or the toilet fill valve), form them into a gentle, even loop, taking care not to kink the tubes.

Installing braided metal supply tubes

Braided metal supply tubes are durable, flexible and easy to work with. If you choose to install them, make sure to get the right length. They are manufactured with integral retainer nuts so they can't be trimmed. Measure the distance from the outlets on the shutoff valves to the faucet tailpieces and purchase braided metal supply tubes that will span that distance comfortably. With the shutoff valves turned off, attach the retaining nuts on the tubes to the outlets on the shut-offs and to the faucet tailpieces to make sure the tubes will fit correctly. If they're too short, the easiest remedy is to return them and replace them with longer tubes. If they're too long, you can either return them or fold the tubes into a gentle loop during installation, as seen in the tip photo to the left.

HOW TO INSTALL BRAIDED METAL SUPPLY TUBES

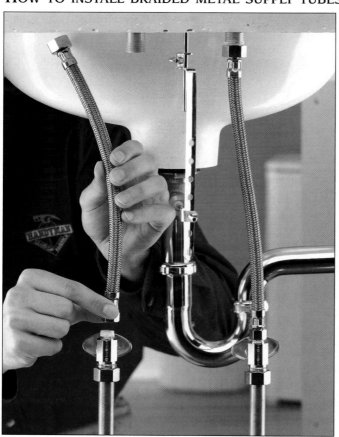

1 Apply pipe joint compound to the female connector and apply Teflon tape to the male connector, then hand tighten the supply tube connection at each shutoff valve.

2 Apply pipe joint compound and Teflon tape to the connectors at the faucet tailpieces, then hand-tighten the tube retainer nut. Use a pliers or adjustable wrench to secure each connection, being careful not to overtighten the nuts. It's very easy to tighten up the connection later if you notice a drip when the supply is turned back on.

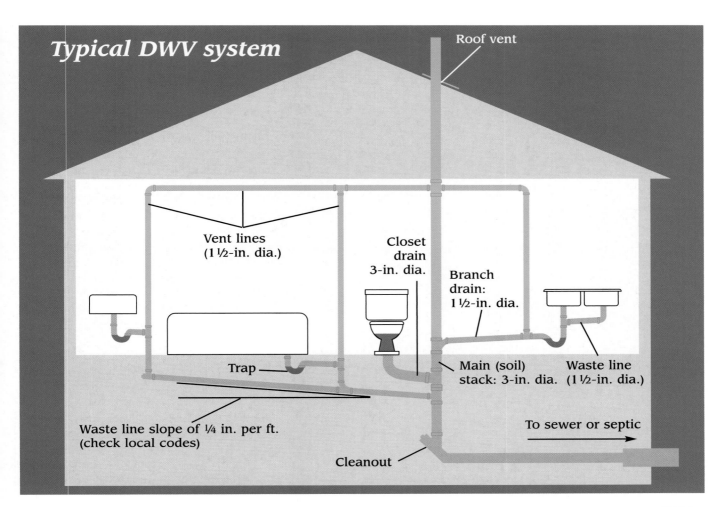

Typical DWV system

Roof vent

Vent lines
(1 ½-in. dia.)

Closet
drain
3-in. dia.

Branch
drain:
1 ½-in. dia.

Trap

Main (soil)
stack: 3-in. dia.

Waste line
(1 ½-in. dia.)

Waste line slope of ¼ in. per ft.
(check local codes)

To sewer or septic

Cleanout

Clearing drain-waste systems

Most drain clogs occur in the fixtures themselves: if you have a clogged drain, that is the first place to start looking. If the drain and trap for your tub, toilet or sink gets a clean bill of health, it's probable that the source of the clog is in the drain waste pipes themselves. While these clogs are more difficult to deal with, the right equipment and a little educated guesswork can allow you to take care of it yourself. The primary tool used to combat drain line clogs is the hand auger, sometimes called a "snake."

DWV leaks are rarer and usually easier to fix than pressurized water supply leaks. Suspect a DWV or fixture leak if water shows

The hand auger, or "snake" is as useful for combatting clogs in drain lines as it is in fixtures, like toilets.

up intermittently, rather that at a constant rate. If you've identified a leaking fitting or pipe, make sure nobody uses appliances and fixtures upstream from the leak while you are performing the repair.

Clearing branch drains. To clean out a clog in a branch drain (a horizontal pipe between the fixture trap and a drain stack), begin at the the connection where the trap arm joins the drain line. Remove the P-trap or whatever fitting allows you to get closest to the wall. Look for an access panel for shower and tub traps. You may need to auger through the drain for showers and through the overflow on tubs, requiring you to remove mechanical drain plug assemblies and drain strainers. For sinks, attempt to remove the trap arm from the wall after removing the trap. Next, push the auger cable into the waste line until it bumps against a bend, then set the auger lock, leaving about 6 in. of cable outside the pipe. Turn the auger handle clockwise to get past the bend. Loosen the lock and push the cable farther into the opening. If you encounter a block you can't get through, set the lock and attempt to bore through the clog by pushing and turning clockwise. If the cable bogs down, turn counter clockwise and withdraw the cable a little until it is free. If you make some progress with continuous resistance, it is probably a soap clog. Repeatedly bore through the clog then wind in the cable by turning clockwise with the lock released. You may tangle with a flexible clog, like hair or cloth. Release the auger lock. Crank clockwise as you pull the obstruction out. Reconnect the trap and flush the drain with hot water.

Clearing drum traps. Older bathtubs, showers, and lavatories (bathroom sinks) may drain into a common iron or lead drum trap. Drum traps are usually located next to the bathtub. A round cover with a bolt-like grip will be flush with the floor or flush with the ceiling below the bathroom. Build a dike of old towels around the drum trap before opening the cover, especially if the fixtures are backed up. If the cover is located in the ceiling, position a large bucket or trash can underneath the trap. Remove the cover and clean out drain lines running toward the fixtures with a hand auger (See augering instructions, above).

Cleanout plugs. Whether on a floor drain, branch drain or drum trap, cleanout plugs may develop leaks. Cleanouts either screw in with threads or they pull out, sometimes after loosening a wing nut that expands the core. Unscrew a threaded plug and thoroughly clean the threads with a small wire brush. Apply generous quantities of pipe joint compound to the threads of a cleanout plug before tightening it back in. Alternatively, measure the diameter of the opening and buy a new expandable rubber plug. Can't remove the old plug? Apply penetrating oil and use a cold chisel and a hammer to attempt to rotate the plug. Or, you can break up the old plug with the hammer and chisel, but be careful not to lose shards down the drain.

Unclogging floor drains

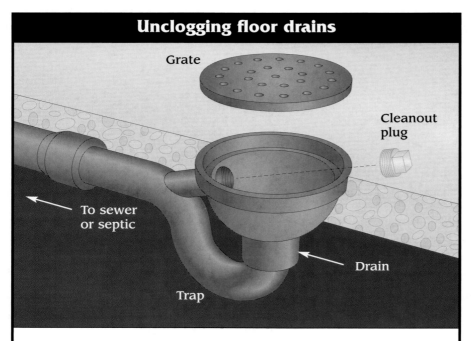

Grate

Cleanout plug

To sewer or septic

Drain

Trap

Not every house has a floor drain in the basement, but if your house has one it's likely that it will need attention from time to time. Most floor drains have a cleanout access covered with a plug that allows you to bypass the drain trap for a better angle at clearing blockages. Simply remove the grate covering the drain, then unscrew the cleanout plug (this normally requires a pipe wrench, perhaps some penetrating oil to loosen the connection, and a bit of elbow grease). Once the plug is out you can run a hand auger into the cleanout and out toward the street.

If the auger fails, another alternative is to use a *blow bag* to clear the blockage. A blow bag is a simple device that attaches to the end of a garden hose. It expands against the inside of the pipe under water pressure, forming a plug, then sends out a jet of water that builds pressure to blast loose the clog. Watch out that the blow bag itself doesn't come shooting back at you like an out-of-control water balloon. Blow bags will not work if the water pressure can escape up a vent or another branch line between the blow bag and the clog. If water is backing out the drain line from other fixtures, then the clog lies past the connections to other fixtures and nearer the street. Find the appropriate cleanout and follow instructions on the next page or call a professional drain cleaning service. Once you successfully clear the drain line, clear the trap next using a hand auger. Wrap Teflon tape clockwise on the threads of the cleanout plug before replacing. Replace the grate.

Clearing clogged main drains

Clearing clogged drain lines can be moderately to extremely difficult, depending on the circumstance. If your house is relatively new and the rough-in plumbers were conscientious, your DWV system will have accessible, easy to remove cleanout plugs. Your clog may be a wayward rag or sponge flushed down a toilet, hopefully removed with a hand auger if you can locate the clog. Alternatively, an older house may have fewer plugs and these may be corroded in place. The clog may be a root that has penetrated an old hub and spigot joint in the sewer. You will never get through a root with a hand auger. You can rent power augers, but these are dangerous and difficult to work with. Below we offer some steps for clearing clogged branch and main drain lines, but remember, your best tool might be the phone book to call a drain cleaning service.

Scenario #1

Your bathroom lavatory (sink) and tub are not draining, but the toilet flushes fine. An examination of the DWV pipes in the space below the bathroom reveals that a branch drain line serving these fixtures joins the main stack below the toilet. You therefore suspect that the clog is in this branch line.

Solution

Position a bucket under the cleanout plug. Remove the plug with a large pipe wrench or adjustable wrench. Clear the clog in the branch line with a hand auger. See pages 51 to 52 for tips on augering techniques. Clean pipe and cap threads with a stiff wire brush. Wrap Teflon tape clockwise onto cap threads before replacing it.

Scenario #2

All the fixtures in the bathroom are not draining, but fixtures and appliances in the kitchen and laundry room are working fine. You believe that none of the functioning fixtures drain into the stack that drains and vents the bathroom fixtures. You therefore suspect that the stack or the section of building drain that drains the stack is plugged. If other drain lines that still work join the segment of horizontal drain line that empties the stack, the plug is probably in the stack.

Solution

Run a hand auger cable down the stack vent above the bathroom. You may need to rent or buy a longer hand auger with a thicker cable if all you have is a short, flimsy do-it-yourself model. Only perform this operation by yourself when it can be done safely. If you encounter the clog, pull about 18 in. of cable out of the drum of the auger and tighten the lock screw. Turn clockwise and push to break through the clog. Repeat this action until you break through the clog. You may need to penetrate the clog more than once to open the drain. Flush the stack afterwards with hot water.

If the clog is below the stack, locate the cleanout plug where the vertical soil stack meets the horizontal building drain. Loosen the plug (you may need to apply penetrating oil). You may need to fill buckets of backed up water from the loosened plug, retightening it in between fills, until you can safely remove the plug. Use a hand auger as described above until clog is cleared. Clean the threads of the cleanout and plug with a stiff wire brush. Apply Teflon tape to the threads of the plug in a clockwise direction or use a non-hardening pipe joint compound, then replace the plug.

Scenario #3

Fixtures in the basement are backing up with water when fixtures on the top two floors are used. You therefore suspect that the large horizontal building drain, the house trap, or the sewer is plugged. These are the most difficult to clear without power equipment. Consider hiring a drain cleaning service.

Solution

Locate the house trap, if there is one. This will appear as two plugs flush with the floor, probably close to the street-side or septic-system side of the house. Loosen, but don't remove, the plug on the street or septic-system side of the trap. If water starts backing around this cap, your clog is in the sewer line beyond the house. You may want to hire a professional drain clearing service at this point. It is unlikely that you will clear a clog in the sewer without a motorized auger. These may be rented but can be dangerous and difficult to operate. Also, roots may have damaged your sewer line.

If no water comes out of the street-side plug, auger the trap. If this doesn't start water flowing, your clog is probably between the trap and the cleanout on the main stack.

Open the house-side trap cover and auger back towards the main stack. You can also attack the clog from the other side by removing the cleanout plug below the stack.

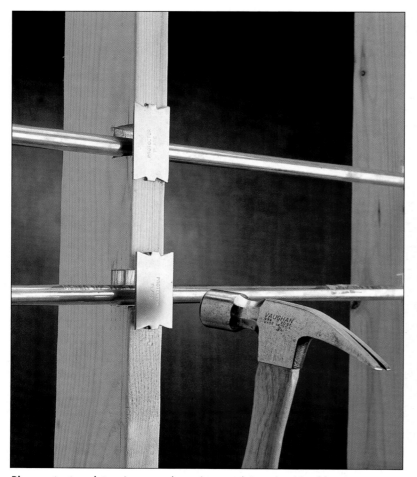

Pipe protector plates. In cases where pipes notch into the side of framing members, protect them with metal pipe protector plates before attaching wall coverings. The plates protect pipes from being pierced later on by nails.

Split, two-hole plastic hangers can be used to suspend copper or plastic pipe. They will not have any negative reactions with the metal, plus they allow the pipe to expand or contract and they don't rattle against metal pipes, as loose metal hangers can. Do not use them if they'll be exposed to direct sunlight or to suspend DWV pipe near a dishwasher or washing machine.

Installing new pipe

Whether you're adding a shower riser or correcting a sag in a DWV branch line, you'll need to work with pipe hangers, clamps, or straps. Keep in mind the "needs" of your pipe. Plastic pipe requires the most support (See chart, next page). A PVC or ABS DWV pipe running from a hot dishwasher or washing machine should be supported along its full length with wood or metal; the heat softens these plastics. Horizontal plastic water pipe should be supported every two or three feet.

Plastic pipe generally should be allowed to move through the hangers without abrasion, since plastic is subject to significant expansion and contraction. For this same reason, flexible plastic tubing should be snaked through stud bays and rigid plastic should generally not be secured immovably at elbows where pipes change direction. When securing DWV or water plastic pipe, stay aware of what happens when the pipe grows or shrinks by a few inches.

If you need to add pipe hangers to water pipe, use a split, two-hole flat strap secured with hex-head bit-tip screws or a split, two-hole, high-eared strap secured with the same. They hold the pipe off the framing, which reduces sound and makes insulating the pipe easier. They also allow the pipe to move lengthwise: an important feature with plastic pipe that can also keep copper hot water pipe from ticking.

If you use metal hangers, use the same metal as the pipe or make sure two unlike metals (like copper and steel) are protected from contacting each other with plastic or insulation. Felted galvanized steel straps insulate against sound, can be used with copper piping, and are appropriate in high stress areas or where plastic hangers will be degraded by sunlight. Pronged wire hangers may back out of joists, so avoid these or use extras to compensate for ones that fail. If running pipe through a stud wall, attach metal protector plates to the wall studs to shield pipes from puncture.

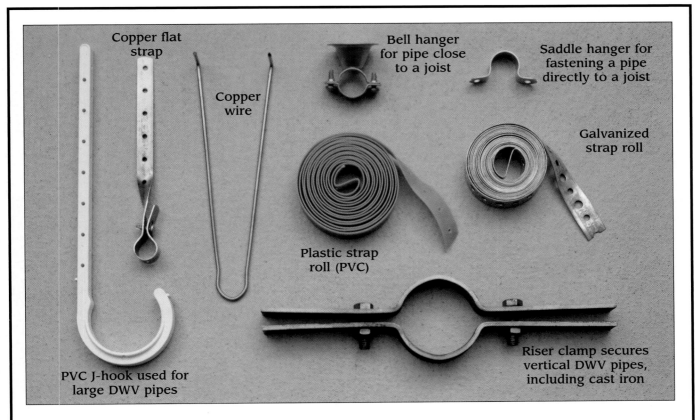

Copper flat strap

Copper wire

Bell hanger for pipe close to a joist

Saddle hanger for fastening a pipe directly to a joist

Galvanized strap roll

Plastic strap roll (PVC)

PVC J-hook used for large DWV pipes

Riser clamp secures vertical DWV pipes, including cast iron

Pipe hangers

Always use special-purpose pipe hangers when running pipe in your home, and make sure the hanger you choose is the best one for the pipe type, size and installation area.

• *Two hole hangers* (See previous page, top right, and the bell and saddle hangers above) can be used with copper, CPVC, PEX, and PB water pipe. These permit hot water pipes to slide during expansion and contraction, prevent pipe rattling, and provide room for pipe insulation. Secure them with screws. Do not use plastic in areas exposed to sunlight. When securing plastic pipe, allow room for expansion and contraction by not securing too close to a change-of-direction fitting.

• *Flat straps* made from solid copper (not plated), or galvanized steel can be used with like-metal pipe in sunlight-exposed areas and where extra strength is needed. Insulated straps may be used with plastic and hot water pipe. Install with No. 6 or 8 stainless steel sheet metal screws. Don't overtighten on hot water pipes, where pipes need to slide lengthwise.

• *Strap rolls* are perforated for ease of use and can be cut to any length you need. They are made from any of the primary materials pipe is made from. Use steel and copper tape with like-metal pipes and plastic tape with plastic pipe or lighter metal pipe. Use heavier galvanized tape on large cast iron drainpipes. Select screws or nails with heads that fill the holes. Consider that plastic drainpipes can bow on long runs, requiring support from above as well as below.

• *Hanger wire* is intended to be driven directly into joists, but has a tendency to loosen or fall out.

• *J-hooks* are used to hang plastic DWV pipe from joists or in walls, where the DWV pipe passes through over-sized holes. Use them to keep the pipe falling at a constant ¼-in.-per-ft. pitch. J-hooks are easy to adjust.

• *Riser clamps* support vertical metal and plastic DWV pipes where they pass through wall plates and subfloors. Additional riser clamps are useful when you need to cut into a stack, leaving the upper portion unsupported. Use with blocking.

Allowable distances between pipe hangers

Material	Support spacing
ABS	4 ft.
Galvanized	12 ft.
Copper	6 ft.
CPVC	3 ft.
Cast iron	5 ft.
PVC	4 ft.
PB pipe	32 in.

Wrap thin fiberglass insulation loosely around water pipes and secure the seams with duct tape. Overlap should be about half the width of the material. Avoid wrapping the pipes too tightly, as that will reduce the insulating value of the fiberglass. This is an inexpensive alternative to using pipe pads (next page).

Insulating & quieting pipes

There are two reasons that insulating your pipes is a good idea: one is to keep them quiet, and the other is to control water temperature (keep hot water hot or keep water in pipes from freezing). Insulating, however, is just one of several ways we'll discuss to quiet noisy pipes.

Temperature control. Insulate cold water pipes in unheated spaces and insulate hot water pipes everywhere. For best results, use pipe pads (extruded closed cell foam insulation—See next page).

Quieting noise. Here is a look at some common causes for irritating pipe noises and some suggested solutions:

• *Water supply-line noise* (water hammer). Water hammer is just what you'd expect: a vibrating pounding that happens when water flow stops suddenly. It can happen when faucets are turned off or, more typically, when washing machines or dishwasher valves shut. Water hammer can eventually cause leaks and cut short the lifespan of valves on appliances, which may be costly to fix. The annoying sound is made worse when rattling pipes rest loosely against pipe hangers or against the framing of your house.

Solutions: You may have noticed a short, capped section of pipe or a chambered device standing on a supply line to a washing machine, sink, or other fixture. This is an *air chamber* (See tip box, next page). If you don't see air chambers near fixtures on your supply lines, consider adding them. When a supply valve or faucet is shut off, air trapped inside the chamber compresses and absorbs the momentum of the flowing water. This prevents the water from "hammering" against your supply lines. You may have air chambers hidden behind walls, or lined up in a manifold where the main line enters the house, or they may never have been installed. Sometimes, air chambers can fill with water and stop working. Draining your supply pipes allows the chambers to drain. This is a good first step to solving your water hammer problems.

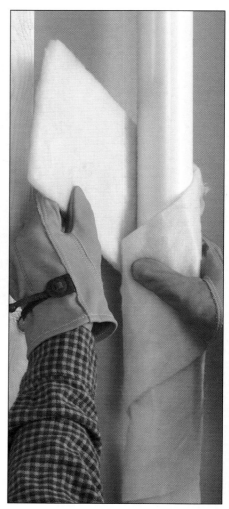

Dampen drain noise by wrapping drain pipes with fiberglass insulation or foam carpet padding. In most cases, this will require you to cut into your walls, so make sure the noise is truly objectionable.

Other possibilities: If you can identify pipes that are rattling from water hammer, check the hangers, especially if they're metal, and add or replace them as needed (See pages 54 to 55) Also check to see if the hot and cold pipes are contacting one another, causing clattering. If so, wrap one (usually the hot) or both lines with pipe pads. If none of these steps work, you may have water pressure that's too high, but can be corrected with a pressure reducing valve (See page 19).

• *DWV-line noise.* In most cases, loud draining sounds can be eliminated by wrapping pipes with fiberglass insulation or foam carpet padding.

TIPS FOR USING PIPE PADS

Hanger wires. Use pipe pads to steady supply pipes in pipe hanger wire. Wrap the section of pipe so the pad extends an inch or two beyond the hanger on each side.

Pre-mitered pads with self-adhesive ends eliminate tricky cutting around elbows and fittings and ensure a tight pad joint.

Use extruded, closed-cell foam pipe pads to cushion rattling pipes in a hanger or against wood. As the water temperature inside the pipes changes, the pipes will lengthen or shorten, making intermittent contact with joists and studs a possibility. Try to stuff some of the padding into cutouts that pipes run through in subfloors, wall and ceilings.

Air chambers

Option 1 (LEFT): Do-it-yourself air chamber. An air chamber can be made from a length of capped copper tubing. Simply install a length of vertical tubing on a horizontal run of piping and cap the end. The vertical tube will hold air and provide a cushion against water hammer.

Option 2 (RIGHT): Ready-made air chamber. Air chambers can also be purchased from plumbing supply stores. These devices can be installed at an angle or upside down.

If your water supply pipes knock or rattle inside the walls, your plumbing system has a condition known as *water hammer.* The hammering is caused by the force of the rushing water within the pipes. When a valve is suddenly closed, the momentum of the rushing water "hammers" against the pipes. The way to alleviate the hammering is to install air chambers, which provide a small cushion of air to relieve the pressure of the rushing water. Air chambers eventually lose their effectiveness as the air is slowly absorbed into the water. The chambers can be recharged by draining your plumbing system.

You can make your own air cushion chambers or buy commercial air chambers. The commercial chambers are more expensive, but never become waterlogged and can be used upside down. Ideally, air chambers should be as close as possible to the appliance or the fixture that's causing the water hammer. Usually the chamber is put on a tee right in front of the shut-off valve to the appliance. See if you can do this first. For convenience, you may need to install the chamber on the nearest accessible strip of horizontal supply pipe to the fixture or appliance.

Thawing frozen pipes

Frozen pipes can cause terrific damage to a house. Most often, the freezing occurs in pipes near or inside exterior walls with inadequate insulation. As the water inside the pipe freezes, it expands and, in time, can cause the pipe to rupture. In this section you'll find some useful tips for thawing frozen pipes, along with some suggestions for preventing them from freezing in the first place.

If you suspect your water pipes may be frozen, proceed as follows:

If water is flowing at all, leave it on. The flowing water will gradually thaw the pipe. If water does not flow, try to determine the location of the ice blockage. Turn on faucets affected by frozen pipes, leaving the water system on. Look for places where lines run close to outside walls or through unheated spaces. A blockage may feel colder than the rest of the pipe.

If you discover a blockage, melt it with a hair dryer, a heat gun on low (metal pipe only), or a propane torch (metal only), starting from the open-faucet side of the block—if you turn the middle of a blockage to steam, you can burst the pipe. Protect flammable surfaces.

An alternative to using a heating device to thaw pipes is to wrap rags around the frozen pipe and apply boiling water to the rags. Melt the block from the faucet side first. Do not use boiling water on plastic.

If the freeze is behind a wall or deep inside a cabinet, aim a space heater or a heat lamp at the wall, being careful not to damage paint or finish on wall.

Preventing pipes from freezing. To prevent freezing of pipes in an occupied house, you must sometimes get creative. If pipes run through cold crawl spaces or through exterior walls, consider insulating the crawl space or blowing loose insulation into the wall. Sillcocks and other outside water lines should be turned off and drained in northern climates. Keeping a cabinet door open on a cold night is sometimes all it takes to prevent a vulnerable pipe hidden inside or behind the cabinet from freezing. Or position a heat lamp at a safe distance from a problem area. Extreme measures include moving water pipes to interior walls. Below are some additional tips for keeping pipes from freezing:

• Insulate pipes in unheated spaces with extruded closed cell foam insulation (See pages 56 to 57). If fire is a concern, wind fiberglass loosely onto pipe in place of foam, which is flammable.

• Electric heat tape can be used in place of insula-

TIPS FOR THAWING FROZEN PIPES

Place a lamp inside cabinets containing water pipes to help prevent pipes from freezing. Sink base cabinets, because they're typically installed at exterior walls, are an often overlooked, and very vulnerable, location for frozen pipe problems.

Use a space heater to gently warm larger areas and fixtures that may be susceptible to damage from more extreme measures. They can also be used to heat pipes concealed by wall surfaces.

tion. Wrap tape around the pipe, leaving a 2-in. gap between spirals. Hold it in place with duct tape. Heat tape will not work if the power goes out.

• On unusually cold days, open sink cupboards on exterior walls, open cabinets and dishwashers that have pipes running through them or through the walls behind them. Aim lamps at spaces containing vulnerable pipes.

• Remove garden hoses and drain water from them in the fall. Shut off water supply to the sillcock. If the valve has a drain cock, open it.

Here are some general tips for winterizing your plumbing in harsh climates, especially If you're leaving a building unattended for a period of time. Even if your house is fully heated, prepare for winter by turning off and draining outdoor plumbing fixtures, sprinkler systems, hoses, and sillcocks.

• Shut off your water at the main shutoff valve or have your water department shut off the water outside your house.

• Once water is shut off, open all faucets, including outside sillcocks.

• Open any drains on valves. Sometimes the main shutoff by the water meter will have a drain cock, as will the outdoor sillcock.

• Flush toilets then pour diluted antifreeze (according to recommendations of brand) into toilets, sinks, tubs, and other fixtures to replace water in traps with antifreeze solution.

• Replace water in main house trap (if one exists) with antifreeze.

• Replace water in washing machine standpipe (drain) with antifreeze.

• Shut off gas to water heater or flip circuit breaker/remove fuse if water heater is electric.

• Drain water from water heater.

• Drain water from all appliances, including water filters and conditioners.

• If you have a pump, drain water from the pressure tank.

• Sprinkler systems and lines to garages and outbuildings should have the water blown out of them with pressurized air. Shut down the drain-and-waste valve in the house. Open faucets and hose bibs at ends of the outdoor lines (sprinklers will be cleared through the sprinkler heads). Attach a compressor hose to the threaded drain cock on the drain-and-waste valve. Use no more than 50 pounds per square inch air pressure to blow the line. Finally, remove plugs from tee fittings in valve boxes for the winter.

Use a hair dryer or heat gun set on low to gradually warm frozen pipes. Start warming at the side of the blockage that's nearer the closets tap (be sure to open the tap).

ALTERNATIVE: Ice blockages can be melted by wrapping the affected pipe in cloth or toweling and pouring very hot water onto the cloth.

Projects & Repairs

Some home plumbing projects and repairs are not time sensitive. If you're planning to replace an old kitchen sink and faucet, for example, you've got plenty of time to mull over your options, shop for bargains and decide on a plan of action. But others, of course, fall into the "emergency" category. A supply pipe bursts. A toilet clogs and overflows. These repairs will not wait until you've had a chance to study up and prepare for them. That's why it's so important to be ready. Along with the previous chapters discussing plumbing tools, materials and skills, get to know the fixtures in your home. Learn how they work. Identify, in advance, which type you have.

This sectionr focuses on fixtures. Installing them, maintaining them, repairing them. You'll find quick fixes that provide temporary solutions to urgent problems, and you'll find comprehensive projects that may be part of your long-term remodeling plans. Sinks, faucets, dishwashers and disposers, bathtubs, showers, toilets, even water heaters and purification systems are discussed in detail.

Whether you're in a state of emergency or midway through a kitchen remodel, knowledge, skill and the correct tools and materials are the keys to coming out of the situation on top. In addition to reading up on the subject, don't be shy about contacting professionals or even your local inspector if you have a question or if the information you find doesn't match your situation exactly. And when working with new plumbing fixtures, always defer to the instructions that accompany the fixture. There is a lot of variation from manufacturer to manufacturer and from model to model. In plumbing, one-size-fits-all is seldom the case.

Draining your water supply system

Before starting any plumbing project or repair involving your water supply system, drain the water from all your supply pipes. Start by turning off the water supply into your home at the shutoff valve located next to the water meter. Then open all the faucet taps, both hot and cold, in the house (or at least the highest and lowest tap).

Remove drain cocks from shutoff valves that have them. This removes any pressurized air from the system. Don't forget to replace the drain cocks before you turn the supply back on at the main shutoff.

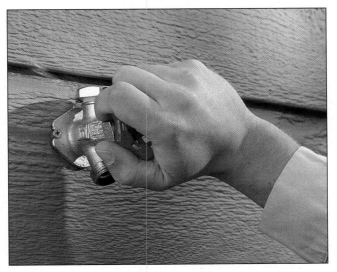

Don't forget about sillcocks. Open all exterior taps completely to drain the water out of their supply lines as well.

Sink Faucets & Basins

Sinks, whether in the kitchen, bathroom, bar or utility room, receive constant use. Consequently, the faucets and basins require regular maintenance and occasional replacement. Any homeowner with even a passing interest in plumbing will almost certainly be confronted with a faucet or sink project at some time.

This chapter shows you how to diagnose and repair common problems you'll encounter with sinks, faucets and basins (including drains). And for those times when a replacement makes more sense than a repair, you'll find information on how to remove your old sink or faucet and replace it with a new fixture.

Because the manner in which you repair or install a faucet depends greatly on the type of faucet, use the information on the next two pages to help you identify which type of faucet you own before attempting repairs.

In most modern homes, hooking up a kitchen sink is complicated by the fact that the sink drain and supply systems are tied into a garbage disposer and dishwasher. Waste water from the dishwasher is discharged through a tube that runs up into an air gap protector mounted to the sink deck to prevent backflow, then runs out from the air gap and into the disposer. From the disposer, the waste water enters a waste pipe that carries it, finally, to the sink drain, where it connects to the pipe with a T-fitting. The only accommodation made for the dishwasher on the supply side is to install a shutoff with two ports in the hot water supply tube: one port leads to the faucet, the other to the dishwasher (See page 102).

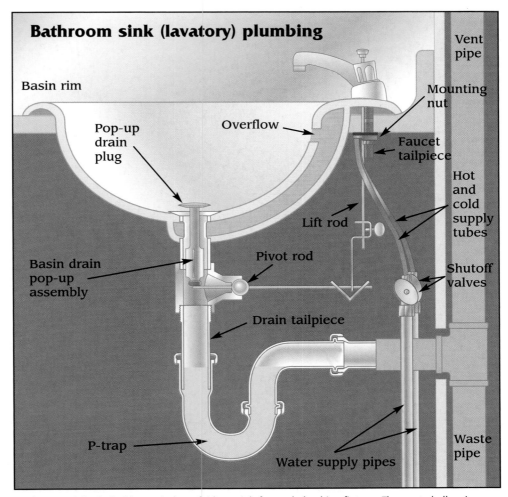

Bathroom sinks (called lavatories) are fairly straightforward plumbing fixtures. The most challenging task when installing a new lavatory and faucet is adjusting the pop-up drain assembly that is usually included with most bathroom faucets sold today.

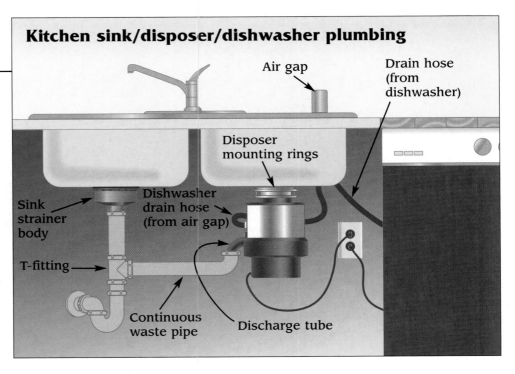

Sink Faucets

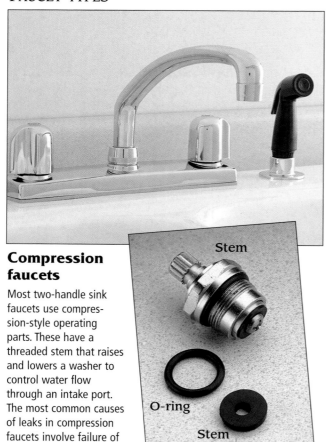

Kitchen and bathroom sink faucets get a lot of attention—on the repair and replacement end, as well as the user end. From a repair standpoint, you'll find that the techniques and troubleshooting solutions have strong similarities, whether you're dealing with a single-handle kitchen faucet or a two-handle lavatory faucet (or a bar-sink faucet, laundry-tub faucet or exterior sillcock, for that matter). When replacing a faucet, of course, you'll want to pay attention to the type of sink the faucet is installed on and make your choice accordingly. Note: For information on tub and shower faucets, refer to Tubs & showers, pages 108 to 133.

Faucets for kitchen and bathroom sinks are selected and replaced for aesthetic reasons as much as any other purpose. Replacing an out-of-date or worn faucet with a newer, more stylish one is a quick and inexpensive way to freshen up a room.

FAUCET TYPES

Compression faucets

Most two-handle sink faucets use compression-style operating parts. These have a threaded stem that raises and lowers a washer to control water flow through an intake port. The most common causes of leaks in compression faucets involve failure of the O-ring or stem washer. See pages 67 to 69.

Ball faucets

Found on single-handle faucets, ball-type mechanisms include a hollow ball that fits into a dome-shaped housing. Rotating the housing and ball causes inlet holes in the ball to align with outlets in the faucet body, letting water flow up through the ball and out the spout. See pages 70 to 71.

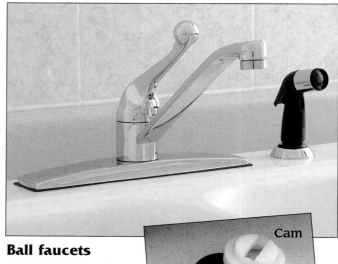

Repairing sink faucets

Leaks and clogs in kitchen and bathroom faucets are generally easy to repair. In most cases, spouts drip and handles leak when washers, rings, and seals inside the faucet fail. These parts are usually made of a black, rubbery material called *neoprene* that eventually wears out. Mineral buildup and insufficiently tightened parts can also cause seals to fail. By far the most common remedy for a leaky faucet is to replace the neoprene parts or make a related repair, such as cleaning, smoothing, tightening or replacing parts that come in contact with the neoprene. Hardware and plumbing supply stores carry generic replacement parts as well as kits for specific brands of faucets. Whenever available, choose repair kits or parts made by the original faucet manufacturer.

The faucet assemblies we display on the following pages feature the most typical valve styles. However, in a century of indoor plumbing, the variations have been almost endless. If you're unable to find specific information on how to repair your sink faucet, the best advice is probably to follow this general sequence: remove and disassemble the faucet parts, laying them out carefully and in the order you removed them; inspect the parts and replace any that show wear; reassemble the parts in the same sequence. Keep in mind that it is often easier (and can be cheaper) to replace old faucets and valves than to spend a lot of time inspecting the parts and trying to locate replacement parts that may or may not be available.

Faucet types. Most of the faucets manufactured and sold today fall into one of four basic categories, based on the mechanical characteristics of the faucet valve component that regulates the flow of water. The four types are *compression, ball, disc* and *cartridge* (See photos, below). While certain brand names often associate with particular valve styles, you will not always know what's under the handle until you take it off. But the general shape and size of the visible parts can provide some good clues, as you can see in the photos below. Keep in mind that some modern faucets utilize features of more than one valve style (in particular, compression-style faucets that include a removable cartridge).

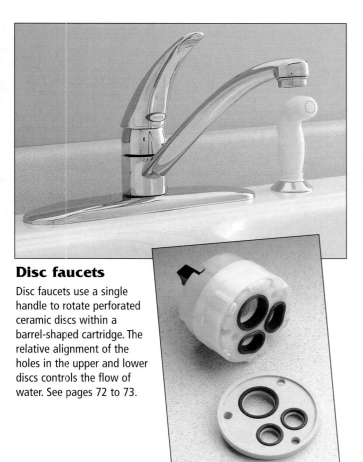

Disc faucets

Disc faucets use a single handle to rotate perforated ceramic discs within a barrel-shaped cartridge. The relative alignment of the holes in the upper and lower discs controls the flow of water. See pages 72 to 73.

Cartridge faucets

Cartridge mechanisms are employed in both single and two-handle faucets. Because the moving parts are all contained within the cartridge, repair is generally a matter of removing and replacing the entire cartridge. See pages 74 to 75.

REMOVING FAUCET HANDLES

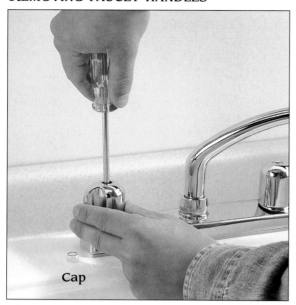

Cap

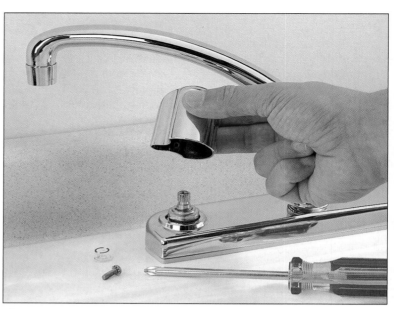

1 **Remove the protective cap** to access handle screws that connect the handles to the stem on the faucet body (make sure to turn the water off at the shutoff valve first). The caps, which often are imprinted with hot and cold labels, usually can be popped out of their housings with a small screwdriver. Store the protective caps in a secure spot while you work—they're very easy to misplace.

2 **Lift the handle off the stem** after removing the screw that secures it. Try gently prying the handle off its stem with a large, tape-protected screwdriver, moving from side to side to rock the handle. Tapping the handle may help. Don't get too aggressive trying to remove the handles: the valve stems are easy to damage. If you cannot remove the handle from the stem, purchase a handle puller (See below). With the handles removed, you'll have access to the stems.

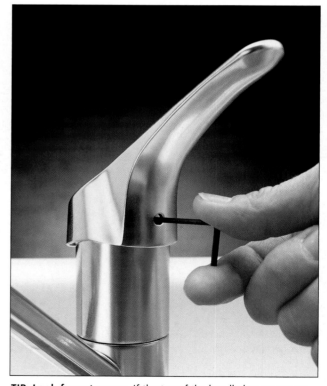

TIP: Use a handle puller. With older compression faucets, it's not uncommon for corrosion to lock the faucet handle onto the faucet stem. You'll need a faucet puller to break this corrosion bond. Simply place the bottom jaws of the puller under the handle and align the puller onto the stem. Then, draw the handle off the stem by rotating the screw.

TIP: Look for setscrews. If the top of the handle has no cap or screw, look for a setscrew at the base of the handle. This may require an Allen wrench to remove. Many single-handle levers are secured with setscrews like the one shown above.

Compression-style faucets

If your sink has separate hot and cold taps, it usually means that you have compression-style faucets. These use replaceable stem washers or diaphragms to block water flow when the faucet is closed. They also employ an O-ring, a packing washer (or, on very old models, packing string) to keep water from leaking out at the base of the faucet when the water is on. In some cases, you will find cartridges instead of valve stems in two handle faucets. These are not compression valves, but since they function the same way they can be treated as if they are.

Troubleshooting tips. If the faucet spout drips when the water is turned off, the stem washer (or its functional equivalent) is probably worn, but you may need to fix or replace a worn valve seat as well. If water leaks out at the base of the handle, you probably need a new O-ring, packing washer, or fresh packing string. Replace the stem washer and O-ring if they appear worn, regardless of the problem you are trying to fix.

Tools for working on cartridge faucets. Three special tools may be necessary. If you cannot remove a handle, obtain a handle puller (See previous page). If replacement stem washers wear out quickly, your valve seat might be damaged. First, try to remove the old valve seat with a seat wrench (See next page) and replace it with a new seat. If it can't be removed, buy a seat-dressing tool to smooth the roughened brass. If all else fails, replace the entire faucet.

Fixing a leak from the base of a handle. Disassemble the faucet handle and bring the stem and the other parts with you to a plumbing supply store. Find replacement parts as needed and reassemble the faucet, applying heatproof grease or petroleum jelly to new

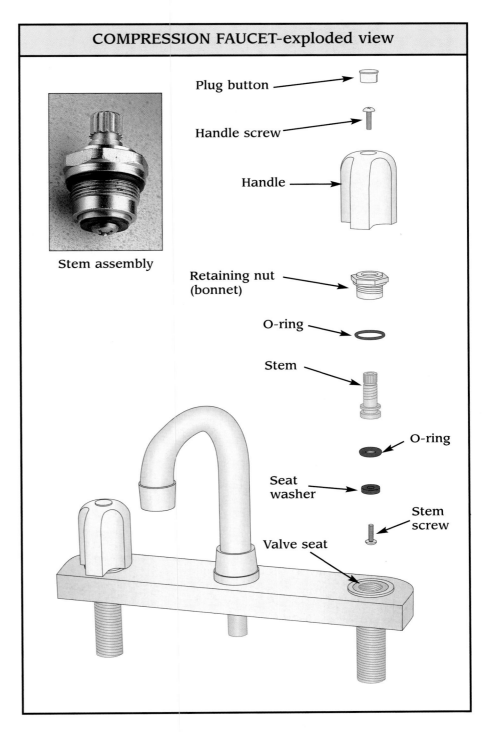

COMPRESSION FAUCET-exploded view

Stem assembly

Plug button

Handle screw

Handle

Retaining nut (bonnet)

O-ring

Stem

O-ring

Seat washer

Stem screw

Valve seat

neoprene parts before installing. If the stem is sealed with packing string, wind new packing string around the stem or push it into the cavity in the packing nut. The string will compress, so apply a little more than enough to fill the space. Tightening the retaining nut slightly after reassembly can correct a newly packed faucet that continues to leak. Packing string

can sometimes substitute for a worn O-ring or a packing washer.

Fixing a dripping spout. The stem washer should be replaced (See page 68) if the spout drips or if the washer is worn. Consider replacing all O-rings and packing washers (or packing string) when replacing stem washers. You may need to resurface the valve seat (See page 68).

Typical types of compression faucet handle stems

Standard stem: A standard stem features a "seat" or "stem" washer attached to the base of the spindle with a stem screw. The seat washer blocks water flow through the valve when the faucet is shut. When the water is on, one or more O-rings, a packing washer, or graphite impregnated packing string (rare today) keeps water from leaking out between the spindle and the retaining nut under the faucet.

Tophat (neoprene diaphragm) stem: A snap-on neoprene diaphragm (shaped like a tophat) replaces the stem washer on this style stem. Snap off the old; snap on the new.

Reverse-pressure stem: Remove the spindle and replace this upside-down washer.

Cartridge: This is not a true compression valve, but it performs the same function. There are no washers to replace with a cartridge. Neoprene rings and seals can be replaced if leaks develop, and the cartridge itself may be replaced.

Valve seats

When spouts on compression-type faucets leak, the problem often is the valve-seat portion of the faucet body, not the stem washer. Once you've removed the handle and valve stem, inspect the condition of the valve seat by running your finger around the rim of the valve seat, or use a flashlight to check for imperfections. It should be smooth and free of pits. If not, replace it (See photos, right). If the inside rim of the seat is rounded rather than faceted for a wrench, you'll need to smooth out the surface in place using a valve-seat grinder (See photo below).

To remove a damaged valve seat, you will need a six-sided valve-seat wrench. Insert the end of the wrench that matches the faceting on the threaded bottom of the valve seat into the seat (top photo). Unscrew the seat counterclockwise to remove it (left photo). Replace the valve seat with an exact duplicate.

Valve-seat grinding tool

To regrind a seat, you'll need a tool called, straightforwardly, a valve seat grinder (also called a dressing tool). After removing the valve stem and any washers or diaphragms covering the valve seat, insert the head of the valve seat grinder into the valve opening so it presses against the valve

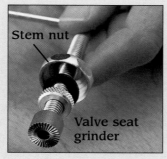

seat. Slip the stem nut from the valve stem over the free end of the grinder. Screw the stem nut onto the male-threaded valve opening. Turn the seat grinder handle three rotations (you don't need to press down—the pressure from the stem nut is sufficient). Remove the grinder and inspect the valve seat with a flashlight. Repeat if necessary, then reassemble the valve.

FIXING A COMPRESSION FAUCET

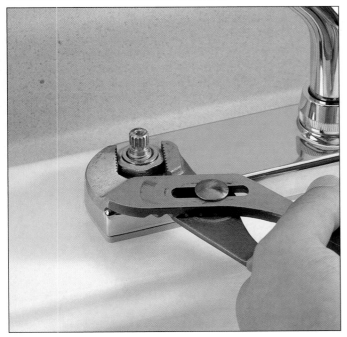

1 **Remove the faucet handles** to gain access to the faucet stems (See page 66)—turn off water first. Then, use slide-jaw pliers to remove the retaining nut from the faucet body. Inspect the valve seats and address problems as needed (See previous page).

2 **Separate the threaded stem spindle** from the retaining nut by unscrewing it.

3 **Remove the stem washer** by unscrewing the stem screw. Typically, a worn washer will have a ring-shaped groove, while a new washer will be flat or cone-shaped on the top. Replace the stem screw and washer. It is helpful to bring the stem with you to a plumbing supply store in order to get the right parts. Tighten the stem washer screw just enough to hold it firmly in place.

4 **Inspect the O-ring** that seals the gap around the base of the spindle. If the O-ring is worn, remove it from the stem with a sharp knife (or, on older models, remove old packing material and packing washers). Also check the O-ring underneath the top flange of the retaining nut. Consider replacing worn rings, packing, and washers at this time too, even if they appear to be in good condition. Apply a coat of heatproof grease or petroleum jelly to new neoprene parts before reassembling the faucet. Once the faucet is reassembled, turn on the water to test it. If you hear a rattling sound when the hot water is running at low flow, you need to tighten the stem washer more. If water leaks from the base of the faucet, try tightening the retaining nut one-quarter turn.

BALL FAUCET-exploded view

Handle

Setscrew

Adjusting ring

Domed cap

Spout

Cam

Cam washer

Ball

Valve seat

Valve seat

Spring

Spring

Valve body

Diverter for sprayer

Ball-style faucets

Ball-style faucets contain a sealed hollow ball with holes in the surface. The faucet handle controls the position of the ball and therefore its holes. When the ball is tilted back, water enters the ball from hot and cold supply-line ports through raised neoprene seals and exits to the hemispherical space beneath the ball, before flowing to the spout. Tilting the ball one way or the other changes the ratio of hot and cold water by changing the size of the openings from the hot and cold ports into the ball.

While the ball rarely needs replacing, neoprene seals and O-rings can become worn, allowing water to drip from the spout, the base of the faucet, or the base of the handle. Good ball-faucet repair kits will include all seals, O-rings, a cam and cam washer, and a spanner wrench.

If the cap or adjusting ring becomes loose, the ball may not form a tight seal with neoprene seals, and water can leak from the spout or spout base. Check to see that the domed faucet cap or adjusting ring is tight before removing the ball and seals.

Leaks from the spout can occur when the springs, which press the valve seals against the ball, become worn. Be sure to replace springs when replacing valve seat seals. Minerals can build up at the supply inlet ports; these should be cleaned.

Before disassembly: First, try stopping a spout leak by gently tightening the domed faucet cap (below the handle) with slide-jaw pliers. If the faucet still leaks, turn off the hot and cold water, open the handle all the way to the right, and remove the handle setscrew. Remove the lever from the faucet. If an adjusting ring is visible, test to see if this is loose with the spanner wrench provided in the repair kit. Tighten a loose ring and turn water back on to see if this stops your leak. Otherwise, leave water off and remove the adjusting ring.

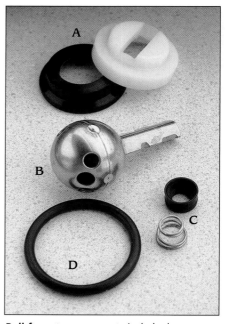

Ball faucet components include the cam and cam washer (A); the ball itself (B); a neoprene valve seal and spring (C) and an O-ring that fits around the valve body cylinder (D).

FIXING A BALL-STYLE FAUCET

1 **Turn the water off** at the shutoff valves. Unscrew the setscrew at the base of the faucet handle and remove the handle. Twist off the cap with slide-jaw pliers—protect the fittings by wrapping the jaws of the pliers with tape. Remove the adjusting ring.

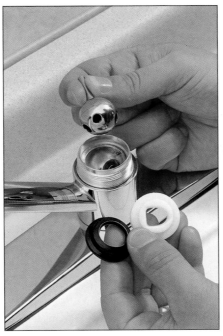

2 **Lift out the cam** and cam washer, then remove the ball and check all parts for wear or damage. A worn ball should be replaced. Metal replacement balls last longer than plastic. Replacement balls often come in kits that include a new cam and washer.

3 **Remove seals and springs** with an awl or a screwdriver. Scrape away mineral deposits in supply inlet ports. Coat replacement seals with heatproof grease and insert new springs.

4 **If your sink has a swivel spout,** twist and pull this off to replace worn O-rings, which can let water leak from the base of the spout. Pry rings off with a pointed tool or cut them off. Coat new rings with heatproof grease, then roll rings down into the grooves.

5 **As long as you have access** to the diverter for the sprayer, inspect the diverter unit and O-ring (if it has one). Replace the diverter or ring as needed (See page 82 for more information).

6 **Reassemble the faucet.** Position the cam over the cam washer so the lug on the side of the cam fits into the notch in the faucet body. Hand-tighten the cap and adjusting ring then test the faucet. If drips occur, tighten the cap with pliers.

DISC FAUCET-exploded view

Handle

Setscrew

Cap

Cartridge screws

Cartridge housing

Spout

Spout seal kit

Seals

Disc

Diverter for sprayer

Valve body

Disc-style faucets

Disc faucets operate using two hard discs with per-fectly flat facing surfaces that contain holes or gaps. The lever on the faucet slides the top disc forward, allowing hot and cold water to enter the mixing cham-ber, which opens to the spout. Tilting the handle to the right increases the channel size for cold water while decreasing the channel size for hot water. At the extreme right position, the hot chan-nel is completely blocked. Tilting the handle to the left does just the opposite.

Disc valve seals can fail from wear or by becoming encrusted with minerals. Seal failure can cause water to leak from the handle area or from the spout. Either way, you'll want to remove the disc cartridge, clean out mineral deposits, and replace the seals. If this doesn't work, replace the disc cartridge.

The mating discs that oper-ate a disc faucet seldom fail, but worn neoprene seals and rings around the disc openings are a frequent cause of leaks.

Leaky disc-faucet handle? Try this first:

Before going to the effort of dis-assembling and repairing the entire disc faucet assembly, simply remove the handle and cap from the faucet (make sure to turn the water off first). Then, tighten each of the mounting screws that secure the cylinder housing to the valve body. Test to see if the leak has stopped. If not, disassemble the faucet and replace any worn parts (See next page).

FIXING A DISC-STYLE FAUCET

1 Turn off both hot and cold water at the shutoff and open the handle all the way. Use the correct type and size screwdriver or Allen wrench to remove the setscrew. Begin an orderly lineup of removed parts. Remove the handle. Gently pry this off if necessary with a large screwdriver protected with tape or a rag. Unscrew or lift off the cap to reveal screws holding the cartridge in place. If screws are loose, tightening them may fix your leak. Otherwise, remove the screws and lift the cartridge from the valve body. You may need to dislodge the cartridge by gripping its stem with locking pliers. Take care not to damage the stem.

2 Remove the three-neoprene seals from the underside of the cartridge. If you need to use a tool to get the seals out, be careful not to scratch the cartridge.

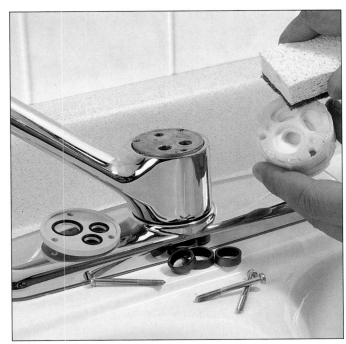

3 Use an abrasive sponge/cloth or a similar non-metallic scrubber to scour away any build-up in or around openings in the discs. Flush out any particles with water. Insert new seals made for your brand and style of faucet, if needed.

4 Also scour around the openings in the faucet body with the scrubber, then assemble the faucet. With the handle in the open position, turn water back on slowly. Close the faucet when water runs steadily. If problems persist, you will have to replace the cartridge.

CARTRIDGE FAUCET-exploded view

Cap

Handle screw

Handle

Domed cap

Pivot retainer

Cartridge screw

Handle adapter & connector assembly

Pivot stop

Retainer nut & bearing washer assembly

Spout

Diverter

D-washer

Cartridge

Retainer clip

Valve body

Sprayer hose

Cartridge-style faucets

Cartridge-style single-handle faucets are made by a number of manufacturers, including *Kohler, Moen,* and *Price-Pfister* (a *Moen* faucet is shown here). Options for fixing them are to clean out deposits, replace neoprene O-rings, or replace the cartridge itself. Cartridge style faucets typically use O-rings to seal against leaks from under the swivel spout and under the handle and to block water flow to the spout when the faucet is off. In short, a worn O-ring can cause any of the usual faucet problems. However, examine the cartridge itself for cracks or corrosion and don't hesitate to replace this if you suspect it's damaged. A word of caution: exercise restraint while tightening cartridge faucets—they tend to be fragile.

Faucet cartridges are hollow tubes that serve as mixing chambers to blend hot and cold water to the desired temperature. The shape and appearance vary widely among manufacturers.

Cartridges, like the one featured here, are used in tub/shower faucets as well as sinks. The general shape and appearance are similar, for the most part, within the product lines of a specific manufacturer. But they do vary quite a bit from manufacturer to manufacturer. As with most faucet parts, the best way to make sure you buy the right cartridge for your faucet is to bring the old cartridge with you to the supply store. It's also a good idea to make a note of the manufacturer and model name or number of the faucet. Since some manufacturers provide a plastic removal tool with new cartridges, you may find it easier to make your purchase based on model name or number and use the tool that's provided with the replacement cartridge to pull the old cartridge.

NOTE: Before disassembling the faucet and removing the cartridge, try simply tightening down the faucet cap to get rid of leaks from the spout. Once you've completed the repair or replacement, remove the aerator and run water through the faucet to dislodge and remove any residue that was loosened during the procedure.

FIXING A CARTRIDGE-STYLE FAUCET

1 **Turn off the water** at the shutoff. Put an old towel in the sink, and prepare to keep track of large and small parts in the order they come off. Use a taped screwdriver as a pry bar to pry off the decorative cover (for most models) over the handle assembly. This may be attached with a carefully-disguised retaining clip that you will need to pull off first with a pair of needle nose pliers. Remove the screw in the handle body. Remove the handle body and lever by tilting the handle back and pulling the handle up. You may be able to remove the spout at this stage. Remove a retainer nut from swivel spout models with tape-protected slide-jaw pliers. Remove the spout. Take off the cover. Be especially careful with chromed plastic covers.

VARIATION Remove a cylinder sleeve on some models to get at the cartridge. Locate and pull out a retainer clip with needle nose pliers to remove the cartridge.

2 **Pull out the cartridge** with slide-jaw pliers. Notice the position of the cartridge ears. If you put it back backwards, hot and cold water will be reversed.

3 **Buy a repair kit** for your model or buy a new cartridge if it appears damaged. Coat replacement O-rings and seals with heat-proof grease or petroleum jelly. Take care not to overtighten fragile plastic parts when reassembling.

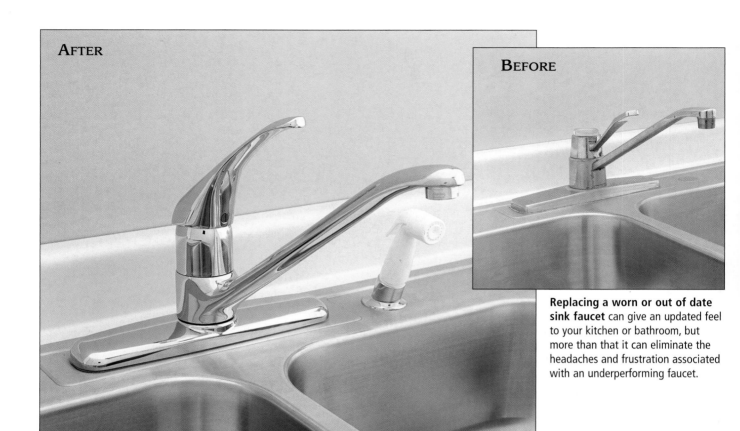

Replacing a worn or out of date sink faucet can give an updated feel to your kitchen or bathroom, but more than that it can eliminate the headaches and frustration associated with an underperforming faucet.

Replacing a deck-mounted faucet

Replacing a sink faucet is perhaps the most common plumbing system update. Because they contain many moving parts and are subject to ongoing use and high pressure, faucets have a limited lifespan, even when well maintained. Add to that the factor of changing styles and design trends among sink faucets, as well as the sheer number of sink faucets found in homes today, and it's easy to see why homeowners can reasonably expect to be confronted with faucet replacement projects on a fairly regular basis.

The good news is that new faucets, while more expensive than one might think, are easy to install. Deck-mounted faucets for the kitchen and bathroom are designed for ease of installation by do-it-yourselfers with only modest plumbing experience.

Look for high quality and reputable brand names when buying replacement faucets. Lower-end units tend to be made with cheaper materials that fail at a much quicker rate. Solid brass under the chrome finish is a good sign, as are solid brass or stainless steel washers and nuts. Avoid handles and faucet bodies made of chromed plastic. These will wear faster than brass, rendering high-use parts such as handles useless. If you do receive plastic or brass-plated steel washers and mounting nuts with your faucet, consider replacing them with solid brass or stainless steel versions of these parts purchased separately.

Deck mounted faucets (also called *center-set faucets*) sit above the sink or counter and attach from the bottom with mounting nuts. Usually, deck-mounted lavatory faucets fit into two holes that are four inches apart on-center, with a third and sometimes smaller hole in the center for the pop-up stop lever rod. Kitchen sinks have a hole under the spout for the sprayer hose attachment and a hole off to the right of the faucet holes for a sprayer handle for a total of four holes. Most faucet valve holes in kitchen sinks are spaced 8 inches apart on-center, for the left and right tailpieces, which usually drop through the first and third holes from the left. Make sure you measure between hole centers on your existing sink before buying a new faucet.

Deck-mounted faucets are sold in single-handle and two-handle models, and you'll generally find your choice of mechanical internal components (ball, cartridge, compression or disk). Faucets with two-handle compression valves are particularly desirable if you have rust flakes in your water.

Sprayer spouts. You can install a faucet with a spout that doubles as a sprayer—the sprayer head is attached to a flexible, retractable hose that feeds up through the spout itself. The dual spout frees the sprayer hole for other purposes, such as a dishwasher air gap, a sink mounted spout for a water filter or hot water dispenser, or a soap dispenser. With one touch of a control button, the spout switches to sprayer mode. This type of faucet is secured from below with one or two mounting nuts, and the flexible copper inlet leads are attached to supply

A deck-mounted faucet with a combination sprayer and spout has a retractable hose attached to a removable spout. The spout converts to a sprayer with the press of a button.

tubes, as you would attach a conventional one-handle faucet. You must provide room below the sink for the long, weighted sprayer hose to retract.

TIP: For a new sink, attach the faucet to the sink before you hang or set the sink. Depending on your maneuvering room, you may also be able to install the drain assembly first.

Water-saving Tip: A fully open faucet will dispense as much as 900 gallons of water per hour. Install a spout-tip shut-off on your kitchen faucet spout to reduce water consumption. This device makes turning water off and on virtually effortless while you're washing vegetables or doing a few dishes. Ask for them at your local home center or plumbing supply store.

HOW TO REPLACE A DECK-MOUNTED FAUCET

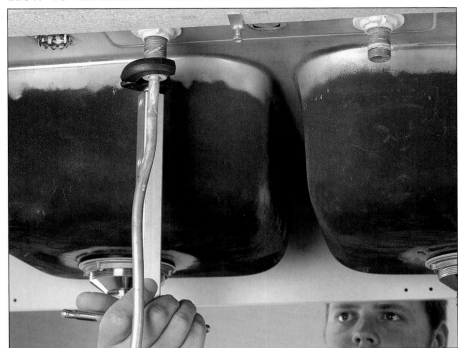

1 **Remove supply nuts** from the tailpieces and loosen the faucet sprayer nipple nut (kitchen sink) to disconnect supply tubes and the sprayer hose. Use slide-jaw pliers if you can, but you may have to purchase a basin wrench. Remove the pop-up rod from a lavatory drain assembly by loosening the thumbscrew on the clevis. Remove mounting nuts from tailpieces and remove faucet body.

2 **Scrape away old plumbers putty** from the sink or counter. Remove old supply tubes from pipe stub outs or shutoff valves. The old tubes will not seal well to the new faucet tailpieces and will need to be replaced. (See pages 44 to 50 for instructions on replacing supply tubes and shutoff valves.)

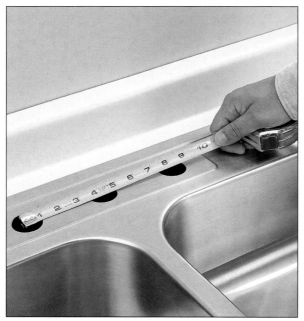

3 **Measure the distance** between the centers of the mounting holes in your kitchen or lavatory sink (usually the first and third holes from the left for a kitchen sink, or the two outside holes for a lavatory). Bring this measurement and the old supply tubes with you to the plumbing supply store when purchasing a replacement faucet.

4 **If the base of your new faucet is flat,** you may apply a thick bead of plumbers putty around the inside of the base. Alternatively, fill the entire cavity with plumber's putty. On cheap faucets, you may have better luck sealing the faucet body by using plumbers putty instead of the plastic mounting plate or foam filler provided with the faucet. The goal is to keep splash water from seeping under the faucet body. If installing a faucet with a pop-up drain, begin hooking up the pop-up assembly before installing the faucet body (See photos, below).

Hooking up a pop-up drain assembly

When installing a lavatory faucet with a pop-up drain assembly, you may find it easier to install the faucet, drain tailpiece and pop-up assembly with the sink disconnected and in an upside-down position. First, insert the lift rod down through its hole behind the spout (left photo). Apply plumbers putty around the faucet base then attach the faucet body by tightening the mounting nuts and washers into the faucet tailpieces. Then, use the thumbscrew to secure the lift rod to the lift strap which attaches to the pop-up assemblies. With the drain closed, the pivot rod should slant up toward the faucet slightly. Adjust the rod/strap attachment until the pop up stop opens and closes properly (right photo). Reinstall the lavatory and complete the faucet hookup as directed in steps 4 through 8 above.

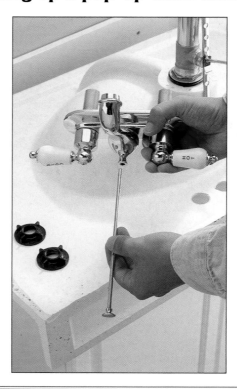

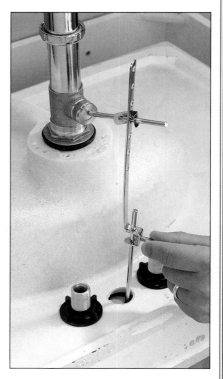

5 (ABOVE) **Twist the friction washers** and mounting nuts over the tailpieces from below the sink and tighten the mounting nut with slide-jaw pliers or a basin wrench. When the faucet is fully tightened and aligned, wipe off any excess putty.

6 (RIGHT) **Connect flexible stainless-steel supply lines** to the shutoff valves and to the tailpieces or copper drop leads, tightening the nuts with a basin wrench. Apply pipe joint compound to the female threads of the supply tubes first, and wrap Teflon tape clockwise on the tailpieces and stop-valve threads. If supply lines are much too long, run them in a loop (See Step 8).

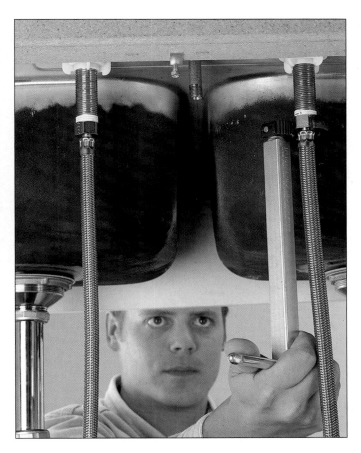

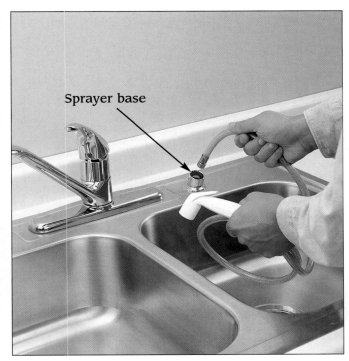

Sprayer base

7 **For a kitchen sink,** attach the sprayer. You may have to punch out a cap in the sink deck from below. Apply plumbers putty to the sprayer base, then secure the base from below with a mounting nut. Thread the hose of the sprayer into the sprayer hole in the sprayer base.

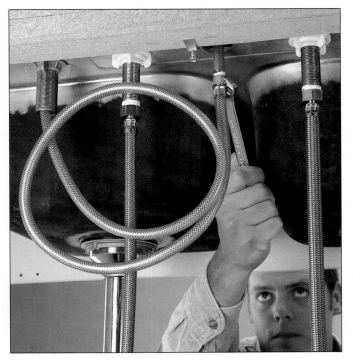

8 **Apply pipe joint compound** to the hose nipple coming out of the faucet body. Thread the sprayer hose onto the nipple. Tighten the nut one-quarter turn beyond hand tight, and then check for leaks and further tighten with a basin wrench if the fitting leaks.

Widespread faucets do not have a faucet body containing the handles and spout. Because the valves are independent of one another, they can fit sink basins with nonstandard hole configurations.

Installing a widespread faucet

Widespread faucets offer greater flexibility than standard deck-mounted faucets, as well as an alternative appearance. Because they are comprised of three independent parts (a spout, a cold-water valve and a hot-water valve) they are not confined to the limitations of an integral base plate and can be installed in virtually unlimited configurations. The valves can be spread up to 16 in. apart, since they are connected beneath the sink with tubing, not fixed to the body of the faucet. Since most sink basins are manufactured with predrilled holes, opting for a widespread faucet is frequently a matter of appearance. But on older basins with nonstandard hole spacing they may be your only option. And it is possible to order sink basins that have not been predrilled so you can drill holes (See page 85) and mount the faucet however you choose.

Removing wide-spread faucets

Existing widespread faucets can be removed by essentially reversing the steps of the installation sequence shown on these pages. Before starting, shut off the water supply at the shutoff valves. First, unthread the manifold connections to the valves and spout tee, then unthread the spout tee from the spout tailpiece and remove the mounting nut and washer. This should allow you to remove the spout. You may need to use a basin wrench. Remove the hot and cold handles then the escutcheon nuts and the escutcheons. Finally, remove the topside mounting nuts and any washers beneath these. This should allow you to pull out the tailpieces from below.

1 Apply a ring of plumbers putty to the flange on the underside of the spout base and slip the threaded shank of the base down through the center hole on the sink. If the spout base has a second threaded member for attaching the spout, make sure both threaded members are aligned according to the manufacturer's instructions (the shank tube should be forward on the *Moen* model shown here). Any O-rings that fit on the top portion of the shank should be in place.

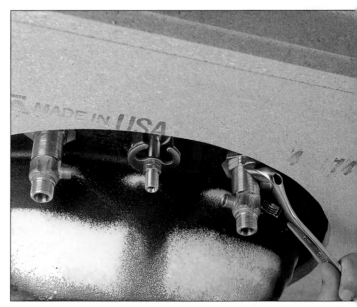

4 From below, hand-tighten the mounting washers that secure the valves against the underside of the sink deck. Make sure the water outlet ports on the valves are aligned so they point at one another. Then, tighten the nuts above and below with a wrench, making certain not to overtighten them.

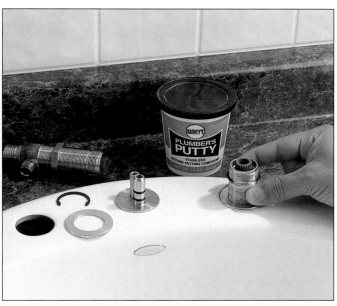

2 **From beneath the sink,** secure the spout base to the lavatory. Specific fittings for this purpose vary by manufacturer. Here, a hole-saw shaped bracket is held against the underside of the lavatory with a retaining nut that threads onto the tailpiece end of the threaded shank. As you tighten the nut with slide-jaw pliers or a basin wrench, you will have to continually adjust the spout to line up with the drain, in order to make a tight connection for the pop-up drain assembly. Work with a helper to make this easier.

3 **Mount the hot and cold valve bodies** to the sink deck, sealing the joints between the valve body flanges and the deck with plumbers putty. The valves typically will be coded for hot (red) and cold (blue) supply connections. The valves are fitted up through the holes in the sink deck and secured from above with retaining clips and mounting nuts. Don't tighten the nuts completely yet. NOTE: You may be required to assemble the valve parts before mounting the valves (See manufacturer's instructions).

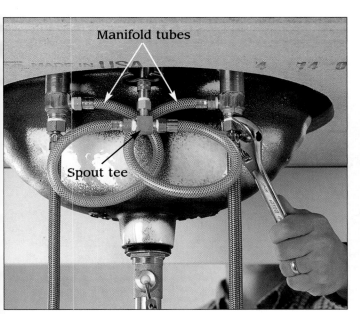

5 **Apply pipe joint compound** to the female threads of the spout tee and wrap three layers of Teflon tape around the threaded tailpiece from the spout base. Connect the tee to the tailpiece. Then, connect the valves to the spout tee with manifold tubes, using pipe joint compound on the female members and teflon tape on the male members. You may have to manipulate the tee to allow for the bow in the manifold tubes. Connect water supply tubes to the valves (but not to the shutoff valves).

6 **Attach the valve handles,** packing plumbers putty underneath the escutcheons to prevent splash water from leaking through the holes. Use valve stem grease on the valve stems where the faucets will sit to keep them from becoming stuck. Applying a drop of Loc-tite to the handle screw will keep the handles from loosening. Follow the manufacturer's instructions to adjust the handles so they are correctly positioned. Hook up the drain pop-up assembly (See pages 94 to 95), then connect the supply tubes to the hot and cold water supply at the shutoff valves.

Maintaining sprayers

If the sprayer head leaks between the handle mount and the sprayer, replace the washer. To remove the sprayer head, unscrew it from the handle mount. Remove the washer and screen at the base of the sprayer and the spray disk at the head of the sprayer. Clean the parts and reassemble.

Sprayer attachment hoses can be troublesome. Frequently, they're made with plastic parts that are less durable than the working parts of the faucet itself. They have a tendency to become clogged. And because the retractable sprayer hoses are subjected to high stress at times, leaks will develop.

Often, a weak spray is caused by a plugged sprayer head. Sprayer heads become clogged at the nozzle or at a screen in the base of the sprayer. Clean these parts as you would clean a spout aerator (See photo, bottom left).

Another problem area is in the diverter valve in the faucet body. The diverter directs water from the faucet to the sprayer hose. A properly working diverter valve shunts all the water from the main spout to the spray head. Improperly working diverters will direct only a portion of the water, or in some cases none at all. Sprayer valves can also lose pressure if the flexible hose from the main spout to the sprayer head becomes kinked. If your sprayer

Replacing the diverter valve in the sink faucet body is a common correction for low-water flow to the sprayer. See pages 70 to 71.

Replace the sprayer. Sometimes the easiest and most effective sprayer repair is simply to replace the entire sprayer and hose assembly. First, disconnect the hose and remove the sprayer. Then, remove the old sprayer base and clean up old plumbers putty. Apply plumbers putty to the bottom of the new sprayer base flange. Insert the tailpiece through the hole and press into place. Finish the installation as shown in steps 7 and 8, page 79.

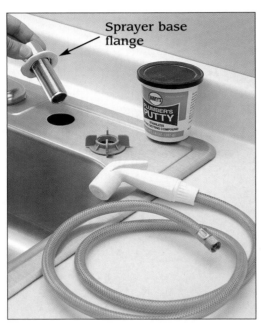

Sprayer base flange

and hose are in good condition, its quite likely that a low-flow problem is caused by a failing diverter valve. To replace the diverter valve (See photo, right) you'll need to disassemble the faucet (See page 29). Once you've gained access to the diverter valve, remove it from the faucet body and clean it with vinegar and a toothbrush. If any wear is noticed, replace the diverter valve and any washers and O-rings, coating new parts with heatproof grease. Reassemble the faucet. Leave the handle open while turning the water on slowly to flush air from the system.

Cleaning aerators & sprayer heads

To remove deposits from a sprayer head or aerator (a spout aerator is shown here), use vinegar, a small brush, and a sharp tool to remove mineral deposits. Reassemble if leaking is not a problem

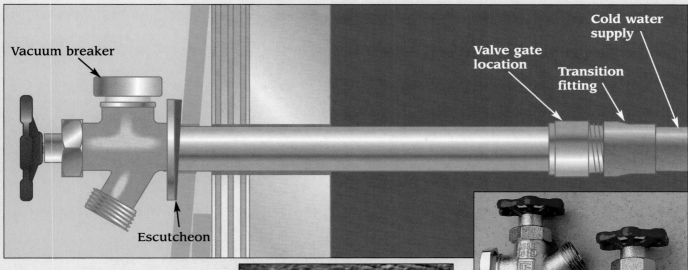

Vacuum breaker

Escutcheon

Valve gate location

Transition fitting

Cold water supply

Installing sillcocks

Faucets that bring water to the exterior of the house are generally called "sillcocks" (but sometimes "hose bibs") because they traditionally are mounted to the mud sill of the house. With direct exterior exposure, they are very vulnerable to freezing and require an anti-freezing mechanism. A frostproof sillcock, designed to resist damage from freezing, has a long shaft and valve stem. Weather parts are left empty when the water is off, since the stem washer and valve seat stop the water in the warmth of your basement. The shaft of the sillcock should slope gently from its connection to the water supply, allowing water to run out.

A sillcock should also include an integral anti-siphon device known as a vacuum breaker. This prevents a hose connected to the sillcock from sucking dirty water into the potable water system, should the water flow backwards in the pipes due to a broken water main.

Using a 1-in. spade bit, drill a hole in a header or rim joist parallel to a cold-water distribution pipe. The hole should be a little below the water pipe, so the shaft of the sillcock will slope toward the spout slightly. Bore clean through the joist, sheathing, and siding.

Insert the sillcock from the outside and mark the location of the

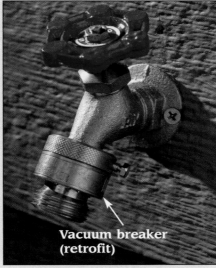

Vacuum breaker (retrofit)

Frost-proof sillcocks. To prevent the water supply leads for sillcocks from freezing and bursting, homeowners in colder climates should install a frost-proof sillcock. These devices are equipped with shutoff gates that are located far enough back into the supply lead that they stop the water flow inside the house. Some have vacuum breakers.

Vacuum breaker required on sillcocks. Vacuum breakers should be installed on all outdoor hose bibs to prevent cross-contamination of your home drinking supply. A broken water main, for example, can sometimes produce backflow. A vacuum breaker prevents backflows. Most building codes now require vacuum breakers in new construction.

screw holes on the siding. Pre drill holes in the siding for screws. Bed the sillcock flange in a thick bead of silicone caulk, and screw the sillcock to the wall. Wipe away excess caulk, and leave the sillcock valve open.

Site down the sillcock across the water pipe or use a straight edge to determine where to tee into the water pipe (See page 28). Attach a tee and leave a short stub of pipe projecting toward the sillcock from the tee. Let the solder cool suffi-

ciently before proceeding.

Wrap Teflon tape clockwise onto the male nipple of the sillcock and screw a female iron pipe (FIP) × copper cup adapter tight onto this. Prepare a ball valve, the stub out from the tee, and another appropriately sized stub for soldering and assemble together as shown. Solder all parts letting the valve cool for ten minutes between soldering each side. Turn on the water when the solder is cool; then close the sillcock.

Photo courtesy of American Standard

Sink basins and lavatories are essentially reservoirs installed between the water supply and the drain system. Installing them typically involves both plumbing and carpentry skills.

Sink Basins

Sink basins themselves require very little maintenance. Refreshing of the caulk seal between the sink rim and counter and backsplash is about the extent of sink basin upkeep, other than regular cleaning. Faucet repair and drain unclogging and repair account for most of the attention you'll typically need to give your sink. But whether it's due to wear or changing design tastes, swapping out your old sink with a fresh new model is a very popular home improvement project.

In this section, you'll find information designed to help you choose and install a new kitchen sink or bathroom lavatory. You'll also find plenty of helpful tips on how to unclog sinks drains, including complex kitchen drains that route through the garbage disposer. Mounting the basin and hooking up the drain are the primary steps involved.

When choosing a new sink or lavatory, variables you'll need to take into account are: *size* (make sure the new sink will fit your existing counter opening, if you wish to keep it intact); *depth* (standard kitchen sinks are 8 inches deep, but many deeper models are available and may well be worth the extra cost); *predrilled hole configuration* (See next page); and *type of material* the sink is made of (See tip box, right).

Sink material options

Stainless steel: Very popular in kitchens because they do not chip, as enameled metal sinks can, and many are very low-priced. Quality is generally a factor of the thickness of the steel. Modern versions are all self-rimming. Available in a brushed "satin" finish or brighter mirror finish. Because they're lightweight, they generally are mounted with retainer clips beneath the countertop.

Enameled steel: Pressed steel sink basins are coated with enamel or porcelain glazing in a selection of color options. Weight depends on the thickness of the metal. Most are self-rimming and easy to install. Finish prone to chipping.

Cast iron: Heavy-weight enameled sink basins made for both kitchens and bathrooms. Due to their weight, retainer clips usually are not needed for self-rimming installation.

Composite: Designed for both kitchens and bathrooms, composite sinks are relatively new on the market. They are lighter than cast iron or solid-surface sinks, but equally thick to deaden sound and provide insulation for hot water. The color is solid throughout, making it easy to disguise scratches.

Solid-surface: Made from the same material as solid-surface countertops, these sinks frequently are rimless (mounted from beneath the countertop) or integral with the countertop.

Vitreous china: Sinks made from vitreous china are relatively lightweight, inexpensive and easy to maintain. They are too fragile for kitchens, however.

Cultured marble: Most bathroom vanity cabinets are fitted with cultured marble basins molded into the countertop surface. Also too fragile for kitchens.

LAVATORY SINK STYLES

Pedestal sinks have "vintage" styling that somehow manages to give bathrooms an updated appearance. The sink basin is supported by the fixed pedestal in

Bathroom vanity cabinets are fitted with an integral sink basin and countertop, usually fashioned from cultured marble.

Wall hung sinks attach to wall studs. With their exposed plumbing they have an appearance that's either "commercial" or "modern," depending on your point of view.

newer models. On older models, the sink is attached to the wall and the pedestal serves only to conceal the plumbing.

Cutting holes in sink decks

Even though virtually all sink basins manufactured and sold in stores today are predrilled with holes for mounting faucets and sprayers, there's still a chance that you'll encounter a situation where it's desirable, if not necessary, to drill new holes into the sink deck. If your kitchen sink faucet has a separate sprayer mounted to the side of the faucet, for example, the standard predrilled holes will be occupied by the faucet and the sprayer. If you want to install a dishwasher, this forces you to choose between eliminating the sprayer in favor of the air gap that's required for dishwasher installation, or drilling a new hole for the air gap. In addition to dishwashers, other sink-mounted appliances are becoming more popular, including water filters and dispensers, liquid soap dispensers and "instant" hot water heaters. You very well may need additional holes in the sink deck to accommodate these. You also may want to order an undrilled sink deck so you can position the holes yourself, exactly where you want them—especially if you're installing a widespread faucet (See pages 80 to 81).

A step bit is used to cut holes into steel sink decks. The holes widen gradually as you drill.

The method and complexity of drilling a hole in a sink deck depends on the material the sink is made of. **Always wear eye and hand protection when drilling a sink deck.**

Stainless steel. The good news is that drilling a faucet hole in sink decks made of stainless steel is a manageable do-it-yourself project. The bad news is that the tool you'll need to drill the hole, called a *step bit*, costs about as much as a low-to-midrange steel

sink. To cut a hole in a stainless steel sink using a step bit, first layout the position of the hole. Then, create an indentation at the centerpoint of the hole with a center punch. Drill a pilot hole all the way through the deck, at the centerpoint indentation (set variable speed drills on low). Chuck the step bit into your drill and drill out the hole. With a step bit, the hole becomes wider as you drill deeper. When you reach the hole outline, stop drilling. Carefully smooth out any rough or sharp edges with a fine metal file.

Porcelain or enamel-glazed steel (not cast iron). Cutting sinks that are glazed with porcelain or enamel over stamped metal is much trickier than drilling an unfinished steel sink. Before drilling, you'll need to remove the enamel or porcelain glazing in the drilling area. There are professional-quality porcelain grinders available for this task, but you can usually get away with using a silicon-carbide grinding stone mounted in a rotary tool (such as a *Dremel*). Remove the glazing down to bare metal, doing your best to stay within or just slightly beyond the hole outline. Then, drill out the metal as with a stainless steel sink (See above).

Cast iron. If you have a cast-iron sink, do not attempt to drill a hole yourself. Look in the telephone directory under "Metal Finishers" to find a professional shop to drill the hole or holes for you.

Cultured marble, composite & solid-surfacing. All of these materials can be drilled with a standard hole saw. For best results, mount the drill in a right-angle drilling guide. Drill on low speed setting with a variable speed drill.

AFTER

BEFORE

Replacing a self-rimming sink

Self-rimming sinks are the most common type installed in kitchens, and are being installed with greater frequency in bathrooms as well. Removing an old sink and replacing it with a new, self-rimming model is a relatively easy job. In addition to the physical installation of the basin into a countertop, shown here, you'll need to install the faucet and basket strainers, attach the drain tailpiece to the sink drain, and install a dishwasher and disposer, if desired. *See illustrations on pages 63 and 102.*

HOW TO REPLACE A SELF-RIMMING SINK

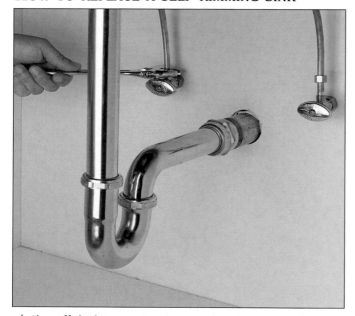

1 **Shut off the house water,** then open faucets and hose bibs to drain water out of the lines. Turn off the hot water heater if the tank will be drawn down. Disconnect the old supply tubes from the shutoff valves.

2 **Loosen the slip nut** holding the basket-strainer tailpiece to the trap. Remove an old disposer, if there is one. Remove the old faucet, if you wish to re-install it on your new sink (See pages 76 to 79), and remove the old basket strainer (See pages 92 to 93) if you wish to reuse it.

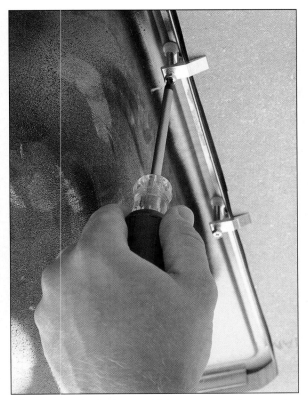

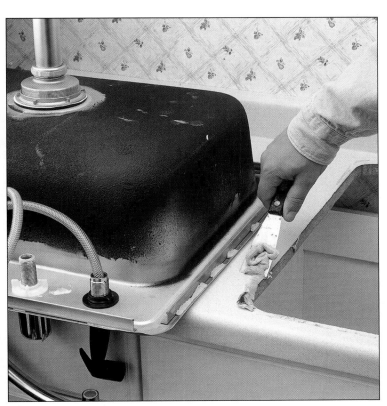

3 **From inside the base cabinet,** unscrew and remove the clips securing the rim of the old sink to the counter. Remove the old sink, cutting caulking seals along the edges of the rim if necessary.

4 **Scrape away old plumber's putty** and caulk and clean up the rim of the counter with grease cutter and a scrubber. Make sure to salvage any sink parts you wish to reuse, then dispose of the old sink.

5 **If the new sink is larger than the old,** trace the outline for the enlarged hole for the sink onto the countertop. Some sink manufacturers provide a template for this purpose. If not, it's easier to trace the profile of the rim onto a piece of cardboard and cut out a template for tracing than it is to invert the sink directly onto the countertop. After tracing the rim outline onto the template, draw a new outline one inch inside the outlet to leave enough countertop surface to support the sink.

6 **Apply layers of masking tape** to the countertop to keep laminate or postform surfaces from flaking off when you cut out the new opening. (if your countertop is made of ceramic tile, score the tiles along the cutting line with a circular saw and masonry blade, carefully chip out the tile in the cutout area, then cut through the subbase with a jig saw.) Cut along the cutting line with a jig saw and fine-tooth blade. If removing large sections of countertop material, be sure to support the cutoff section from below before cutting.

HOW TO REPLACE A SELF-RIMMING SINK

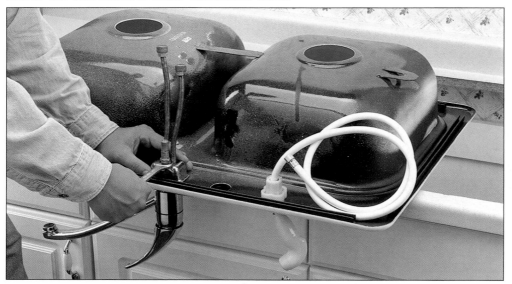

7 Make as many hookups to the sink as possible before installing the basin—it's much easier to reach working with the basin upside down than with your body upside down and laying in a base cabinet.
Kitchen sinks: At the very least, you should be able to hook up the faucet, supply tubes and sprayer (See pages 76 to 82). You should be able to install the baskets strainers (See pages 92 to 93) as well. Another possibility is the garbage disposer (See pages 102 to 103).
Lavatories: Install the faucet (See pages 76 to 82) and the pop-up drain (See pages 94 to 95).

8 After the hookups are finished, place a thick snake of plumbers putty around the rim of the new sink (unless the manufacturer's installation instructions call for silicon caulk). Place the sink in the opening, bedding the rim into the putty evenly. Lighter-weight self-rimming sinks require that retainer clips be installed every 6 inches, from the underside of the countertop (See step 3). Tighten the clip screws until the sink is resting flush on the countertop. Plumber's putty should squeeze out all the way around the rim of the sink. Scrape and wipe off excess plumber's putty

9 Finish making the hookups for the drain, supply tubes and dishwasher/disposer (if needed). After all plumbing hookups are made, open the shutoff valve for the sink and make sure the faucet is open. Then, turn the water back on at the house shutoff (having the system open when water flow is re-established decreases pressure build-up and makes it less likely to trap air in the supply lines). Observe the faucet and drain to make sure there are no leaks.

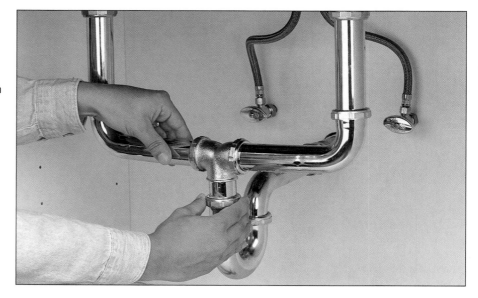

Bathroom vanities

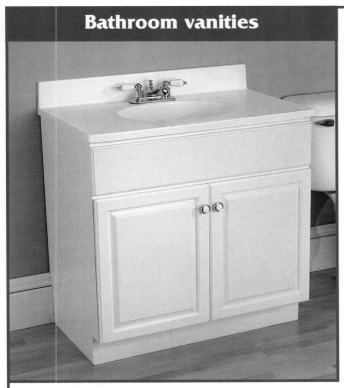

Vanity cabinets are popular for their storage capacity and the design options they offer. Typically, they are equipped with integral countertop/lavatories molded from cultured marble, or with postform countertops and drop-in, self-rimming lavatories.

1 **Remove base molding** in the installation area. Cut holes in the cabinet back (if it has one) or bottom to accommodate the supply and drain pipes entering from the wall or floor. Set the vanity cabinet in position and insert shims between the cabinet and floor, if needed, to level and plumb the cabinet. The cabinet should be arranged so the drain tailpiece from the lavatory will align with the drain arm in the wall or floor.

2 **With the cabinet level and plumb,** attach it to the wall by driving screws (we used 3 in. drywall screws) through the hang strip at the top of the cabinet back and into wall framing members. Cut the shims flush with the edges of the cabinet.

3 **Attach the sink faucet** to the vanity top (See pages 76 to 79 or 80 to 81), as well as the drain and pop-up drain assembly (See pages 95 to 95). Dryset the countertop on the cabinet to make sure the drain components align. Adjust their position as needed. Then, apply a bead of caulk around the tops of the cabinet edges and seat the lavatory into the caulk. Make the final supply and drain hookups, then turn the water supply back on. Trim baseboards to fit against the sides of the cabinet and reinstall them. If gaps between the floor and cabinet are visible, conceal them with quarter-round molding or base shoe.

Wall-hung sinks

1 **Measure up on the wall** from the center of the drain arm (31 in. is standard height for a lavatory deck). For renovation work, you'll need to cut a small hole in the wall covering to create access for installing blocking between wall studs to support the sink hanger hardware. The front face of the blocking should be flush with the back surface of the wallcovering.

Wall-hung sinks are well-suited for smaller bathrooms because they do not occupy any floorspace.

2 **Patch the hole** in the wall. Attach the sink hanging hardware (provided with the sink) to the wall so the basin deck will be 31 in. above the floor and the drain tailpiece will be aligned with the drain arm in the wall. Use lag screws driven into the blocking to attach the hanger, making sure it is level.

3 **Mount the faucet** (See pages 76 to 79) to the sink deck. Then, lower the sink onto the hanger, centering it form side to side. The tabs on the hanger should fit inside the flange on the back of the basin. Once the basin is in place, use a pencil to mark drilling points on the wall through the screw guide holes cast into the back edge of the sink. Remove the sink and drill holes for screw insert sleeves. Drive the sleeves into the holes then rehang the sink. Drive screws with washers through the guide holes and into the sleeves to secure the basin. Hook up the plumbing and caulk between the back edge of the sink and the wall.

Pedestal sinks

Pedestal sinks are supported by the pedestal and with lag screws driven into the wall framing members.

3 **Undo the loose plumbing connections** and remove the sink (lay it carefully on its back lange to prevent damage to the supply tubes or tailpiece). Drill pilot holes for the screws (usually included with the sink) that secure the pedestal to the floor and the basin to the wall. If the holes for the sink screws do not fall over wall framing members, install blocking inside the wall (See Step 1 of "Wall-hung Sinks").

1 Mount the faucet on the sink deck (See pages 76 to 79) and attach the pop-up drain assembly (See pages 94 to 95), including the drain tailpiece. Also attach the drain trap to the drain arm with slip nuts. Set the pedestal on the floor, centered in front of the drain arm in the wall, then place the basin carefully onto the pedestal. Adjust the pedestal so the basin is flush against the wall, then outline the position of the pedestal on the floor with tape. Also mark the position of the screw guide holes in the base onto the floor.

2 Mark and cut supply tubes and the tailpiece as needed to align with the drain trap and the shutoff valves— remove the sink basin before doing any cutting. Once all pipes are cut to length, make the connections (loosely) to be sure they'll work out in the final installation (access to the plumbing is impeded by the pedestal, making it next to impossible to make adjustments once the basin is installed). Before final removal, mark drilling points at the guide holes for securing the basin to the wall.

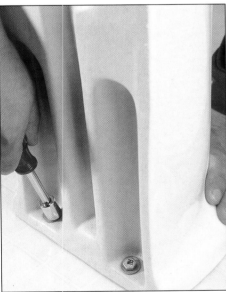

4 Apply silicon caulk to the bottom edges of the pedestal then set it back into position on the floor. Use a nut driver to drive the lag screws through the guide holes in the pedestal base and into the pilot holes in the floor. Don't forget to slip washers onto the screws before inserting them into the guide holes. Take care not to overtighten the screws, as too much pressure will crack the porcelain pedestal base.

5 Apply silicon caulk to the back flange of the sink basin and set it back onto the pedestal. This can be a little tricky, since you need to insert the drain tailpiece into the trap arm as you lower the basin (don't forget to have your slip nuts threaded onto the correct parts first). Once the basin is in position, drive lag screws through the basin flange and into the pilot holes in the wall (See inset photo). Make final plumbing hookups.

BASKET STRAINER
-exploded view

Strainer/stopper

Strainer body

Rubber gasket

Flat washer

Lock nut

Slip-nut washer

Slip-nut

Tailpiece

1 **Use slide-jaw pliers** to loosen the slip nuts above the trap and below the basket strainer. Remove the tailpiece. Loosen the large lock nut holding the basket strainer body to the sink. You may use extra-large channel-lock pliers or a spud wrench or attempt to tap the nut loose with a screwdriver and hammer.

Basket Strainers

Basket strainers serve the double purpose of plugging the kitchen sink and straining food wastes when draining the sink. If you can't keep the basin filled with water, replace the strainer/stopper basket.

If the basket strainer housing leaks where it joins the sink, remove the strainer body and replace the plumber's putty seal between the strainer body and the sink. Replace the strainer gasket and washer beneath the sink at this time also. Replace the lock nut, if it is cracked or corroded.

If the strainer body leaks where it joins the tailpiece, tighten the slip nut or replace the washer beneath the slip nut. Replace a broken slip nut.

If the basket strainer body is damaged or you cannot find replacement parts, replace it.

TIP: Not sure where the leak is coming from? Fill the sink with water and dry it well underneath. If water shows up again, the leak is between the basket strainer body and the sink. Remove the basket strainer and reapply the plumber's putty seal between the basket strainer body and the sink.

TIP: It may be necessary to hold (or have somebody else hold) the strainer body steady from the top as you remove the large lock nut from below. Special basket strainer wrenches are made for this purpose, or you can insert the handles of a pair of pliers into the strainer and steady them with a large screwdriver.

3 **Press a ½ in. thick plumbers-putty snake** around the opening for the rim of the strainer body. Apply pipe joint compound to all threads of the basket-strainer body. Set the strainer body into the drain hole. Position the strainer so the brand name is right side up. Apply pipe-joint compound beneath the rim of the drain hole from under the sink. Slide the gasket and the flat washer over the strainer body and secure them with the lock nut. To tighten the lock nut, hold (or have somebody hold) the strainer body steady from the top with a basket strainer wrench or pliers and screwdriver. Wait ten minutes for the rubber gasket to compress, then try tightening the basket a little more.

2 **With the nut and all washers** removed from the strainer body, knock out the strainer from below. Scrape off old putty and the clean rim of the drain hole with mineral spirits and a scrubbing pad.

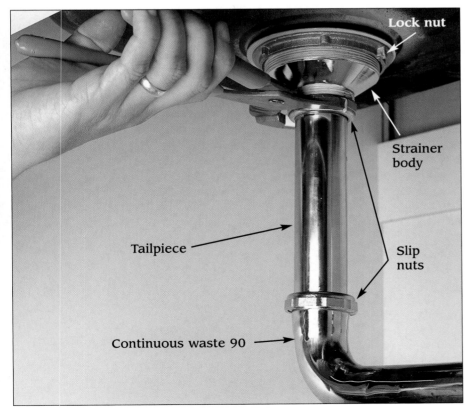

4 **Slip the slip-nut washer** and the slip nut up the tailpiece and hand tighten the tailpiece to the strainer body outlet. Slide the trap slip nut and the washer (bevel facing down) onto the tailpiece. Apply Teflon tape clockwise to the P-trap or the J-bend threads. Hand tighten the J-bend of the continuous waste 90 to the tailpiece with the slip nut. Tighten the slip nut with slide-jaw pliers. You may pad the threads of the pliers with masking tape to avoid scratching chrome-plated brass. Test for leaks.

Lock nut

Strainer body

Tailpiece

Slip nuts

Continuous waste 90

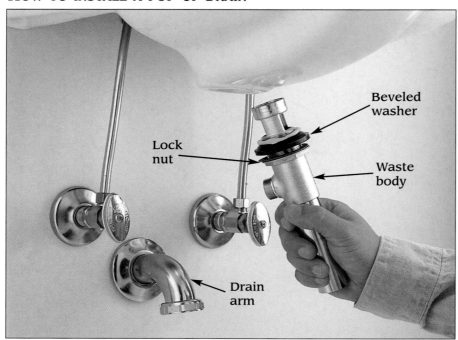

1 **Remove the pop-up stopper** from the new waste body. Unscrew the flange. Screw the lock nut well down the shaft of the waste body, and pull the metal friction washer and rubber gasket after it. The beveled side of the gasket should face up and will press against the bottom of the basin. Apply a plumbers putty snake to the top of the beveled washer and apply pipe joint compound to the female threads of the waste body.

2 **Make a snake of plumber's putty** and press it around the bottom of the flange. Insert the waste body through the drain hole from below. Thread the flange onto the waste body from above.

Installing pop-up drains

Lavatory sinks are usually plugged with a pop-up stopper that's lifted and lowered with a knob on a lift rod behind the spout (See page 63 for an illustration). The heart of this mechanism is the waste body, a hollow tube that contains the pop-up stopper. The waste body has openings where it crosses the overflow channel under the basin. Water flowing down through the overflow channel can enter the waste through these openings.

The waste body screws into a flange that rests on the rim of the hole in the basin. Plumbers putty forms a seal between the basin and the flange. If this seal leaks, water held in the basin will escape to the overflow channel beneath the basin and into the waste. For this reason, two seals must be checked when a basin doesn't hold water: the stopper seal and the flange seal. The waste body penetrates the outer wall of the overflow channel where it drops below the sink. A rubber waste gasket, pressed against the underside of the basin by a friction washer and a lock nut, prevents water from seeping out. If water is leaking from under the sink, this is one of the seals that may have failed. A tailpiece threads to the bottom of the waste body and is held in the trap with a slip nut.

The stopper itself is on a post that is pushed up and let down by a pivot rod. The pivot rod runs through a pivot ball that sits behind the waste body. A beveled nylon washer keeps water from leaking out around the pivot ball. The pivot rod is levered up and down by a lift strap (immediately above the pivot rod) that is pushed up and down by the user with the lift rod.

A pop-up stopper that doesn't seem to seat itself fully in the drain opening or open far enough may need adjusting, or it may need cleaning. You might as well do both at the same time.

Removing an old pop-up drain

Remove the cap holding in the ball and pivot rod and take out the rod. Loosen the slip nut holding the tailpiece in the trap and unthread the tailpiece from the pop-up body. Loosen the lock nut on the underside of the basin with slide jaw pliers, and thread down the nut until you can push the flange up into the basin. Grab the flange with tape-protected pliers and unscrew the flange from the pop-up body. Thoroughly clean old plumbers' putty from the basin.

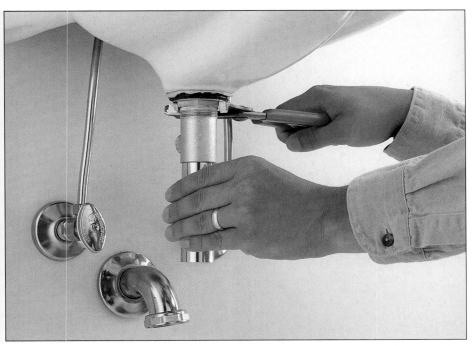

3 **Pull the waste body down** to seat the flange in the basin, then push the gasket and washer up under the lavatory and screw the retaining lock nut on after them. Tighten the lock nut securely with slide jaw pliers, but do not press the gasket out from under the retaining nut.

4 **Put the stopper into the waste.** If there is a hole at the bottom of the stopper post, orient this toward the rear of the lavatory.

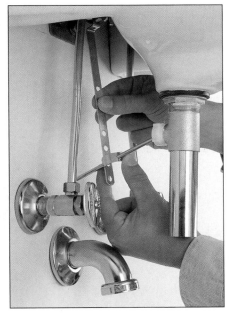

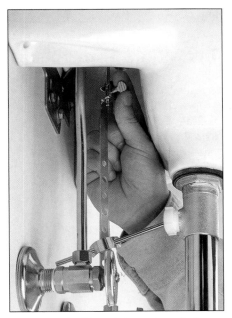

5 **Remove the cap holding the ball** and pivot rod in the waste body and take out the rod, ball, and nylon washer. Spread pipe joint compound in the opening for the ball. Put the nylon washer back in, with the bevel facing where the ball will seat. Wrap Teflon tape clockwise onto the waste body threads for the ball cap, and replace the rod, ball, and cap. The arm should thread into the hole at the base of the stopper post.

6 **Check to see that you can operate** the pop-up stopper with the pivot rod before attaching the pivot rod to the lift strap. You may have to remove the stopper and pivot rod to adjust their connection. When the stopper is working well, pull the pivot rod down, opening the pop-up stopper fully. Attach the lift strap to the pivot rod by inserting the rod in a hole in the strap while sandwiching the spring clip over the hole.

7 **Insert the lift rod down** through the faucet body. Position the lift strap on the pivot rod so the lift strap is aligned under the lift rod. Slide the lift rod into the bracket at the top of the strap behind a thumbscrew. Adjust so that when the knob is fully depressed, the pivot rod is depressed all the way. Tighten the thumbscrew. If the lift strap runs into the trap arm, trim the lift strap.

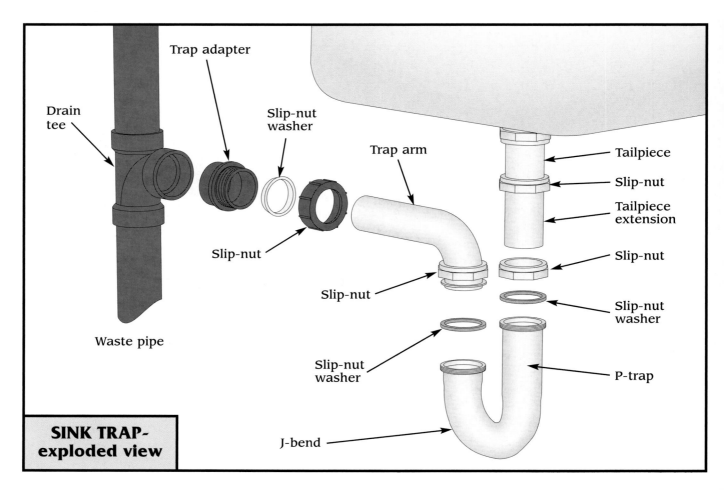

Drain tee

Trap adapter

Slip-nut washer

Trap arm

Tailpiece

Slip-nut

Tailpiece extension

Slip-nut

Slip-nut

Slip-nut washer

Waste pipe

Slip-nut washer

P-trap

J-bend

SINK TRAP- exploded view

Sink traps

Replace sink and lavatory traps that leak or become corroded. In most areas, you can use ABS (a black plastic), PVC (a white plastic), or chromed brass trap fittings. Some localities permit only metal traps since plastic traps will burn in a fire and give off toxic fumes. Because sinks and lavatories are easily accessible, code permits you to join the components of your trap system with slip nuts. This also allows you to easily take apart the trap to clear out clogs.

Lavatories generally have 1¼ in. trap systems and kitchen sinks generally have 1½ in. trap systems. Double lavatories move up to a 1½ in. trap size, but double sinks stay at 1½ in. If you are wondering what size trap you are replacing, measure the outside diameter of the sink tailpiece or trap arm (See photo, right). Tubular trap pieces are sized according to their outside diameters. This allows you to insert the trap arm inside the pipe stub in the wall, since DWV pipes are sized by their inside diameters.

If you will be replacing a sink tailpiece and the trap arm, measure the length of these so you know what the minimum size is that you must get. It's okay to buy these pieces long since they can be cut back, but make sure you get them long enough to span the gaps,

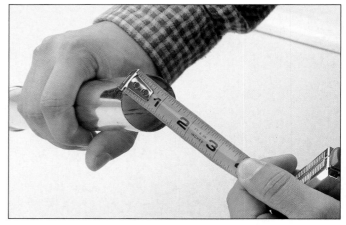

Measure outside diameter to determine which size trap pipes you need for replacement (1¼ and 1½ in. are the standards).

including the overlaps. You want the tailpiece of the sink or lavatory to penetrate the trap as much as possible. You also want close-to-maximum penetration of the trap arm into the trap adapter and stub in the wall.

Here we describe how to measure for trap arms, continuous wastes, and tailpieces. This will be useful if you're changing sinks. If you're simply replacing trap and waste components, take measurements off the old parts to cut your new parts.

Chromed brass vs PVC sink trap fittings

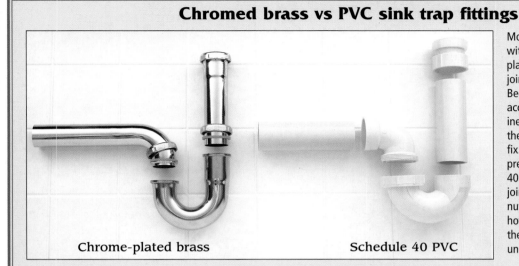

Chrome-plated brass   Schedule 40 PVC

Most homeowners prefer to work with chromed or lighter-weight plastic sink trap fittings that are joined with slip nuts and washers. Because they can be easily accessed, the occasional (and inevitable) leaks that develop when the slip nuts slip are not difficult to fix. But many professional plumbers prefer to use heavy-duty schedule 40 PVC trap components that are joined by solvent-welding, not slip nuts (the J-bend to trap-arm joint, however, is made with a slip nut so the trap can be disconnected and unclogged if needed).

*Important: Not all sink traps are tubular size. Schedule 40 plastic traps and some metal traps are pipe sized, which means the nominal diameter is the same or smaller than the inside diameter. Solvent-welded Schedule 40 plastic is often used with disposers so the joints won't be shaken apart. With pipe-sized traps, a trap adapter or Mission coupling must be used on the sink side of the trap to transition to the tube-sized tailpiece.

Before starting work on your trap system, check with your local building inspector to determine if you will need two traps for a double lavatory (this usually depends on how far the most distant sink is from the waste stub-out in the wall.) Also check to see if you will need a baffle in the tee if you are using a single trap. (A baffle keeps waste from one sink from flowing into the other sink.) Finally, see what types of materials are allowed for building the trap system. Some local codes restrict the use of plastic drain piping outside the wall. Note that a stub-out from the wall must have at least a 1 1/2 in. inside diameter to receive waste from a double lavatory.

Double sinks and lavatories. If one sink is more or less in front of the waste stub-out in the wall, then you will want to install an *end outlet tee* (See page 99). The line up doesn't need to be perfect since you may swing the J-bend on the P-trap one way or another and still have the trap arm approach the wall square. You will attach a trap to the one sink in front of the stub out as you would trap a single sink, except the tailpiece of the sink will be interrupted with a tee. A long pipe with an elbow on the end will run from the side inlet of this tee over to the other basin. This long pipe is called a *continuous waste 90* since it's continuous with the other waste and has a right angle on the end. If the drain stub-out is between the two sinks, then you will want to use a *center outlet tee* (See page 99). With a center

Sink trap materials

Sink trap components are made from a number of materials. Some local codes place restrictions on which types may be used, so be sure to check with your building department if you're not sure. Perhaps the most common material today, particularly when replacing a trap with a store-bought kit, is lightweight plastic (PVC) tubing (A). It is inexpensive and easy to cut and make connections with. But it is also the least durable and the connections are the most prone to coming apart. Chromed brass tubes (B) are also very popular among do-it-yourselfers. Like plastic tubing, chromed tubes are fitted together with slip nuts and are relatively easy to cut (although more touchy than plastic). Chromed tubes also are prone to corrosion and deterioration. Schedule 40 ABS pipe (C) and Schedule 40 PVC (D) are very durable and are joined with tough solvent-welded connections.

outlet, both sinks attach to the same tee with continuous waste 90s.

TIP: When replacing your sink trap, buy quality materials for longer life. If you are using chromed brass trap parts, choose heavier, 17-gauge brass over thinner 20 or higher gauge material. Replace plated steel slip nuts that come with trap kits with solid brass nuts.

Tailpiece extensions

Often, the tailpiece from a pop-up stop waste or a basket strainer isn't long enough to reach the trap. To solve the problem, purchase a longer tailpiece for the basket strainer or pop up stop, or buy a tailpiece extension. The extension on the left in this photo (A) is a slip-joint extension that is simply added to the original tailpiece with a slip nut. The fitting on the right (B) is a 12-in.-long, threaded, extension that screws into the drain body to replace the original tailpiece.

Transition fittings/trap adapters

You may already have a trap adapter with a slip nut on the stub-out in your wall. If not, you can purchase the appropriate fitting if you measure and/or identify the pipes to be joined. Identify nominal DWV pipe sizes by their larger-than-nominal outside diameters. Tubular sizes, used with most trap parts, are equal to the outside diameter.
(A) This trap adapter transitions from a 1½ in. tubular chromed brass trap arm to a 1½ in. threaded galvanized steel pipe. (B) This trap adapter transitions from a 1½ in. tubular PVC trap arm to a 1½ in. Schedule 40 PVC pipe. (C) This solvent-welded adapter transitions from 1½ in. Schedule 40 PVC to 2 in. Schedule 40 PVC. (D) Mission coupling to join any narrower trap arm to a wider stub-out.

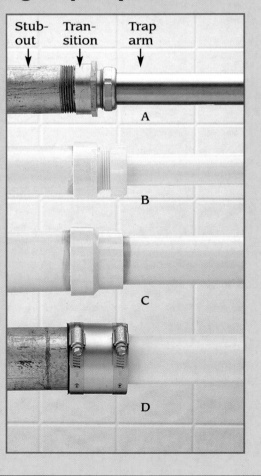

Replacing an end-outlet trap

Replacing a sink trap in an end-outlet configuration (bottom illustration, next page) is very similar to replacing one in the more common center configuration (top illustration). If you have an end-outlet configuration, read the following description as well as pages 100 to 101. End outlet traps work similarly to center outlet wastes except you'll need only one continuous waste arm, and instead of a center outlet tee you'll use a 1½ in. end-outlet slip joint tee.

To install an end outlet trap, first remove the old trap by loosening the slip nuts at the trap adapter and the continuous waste 90 and pulling the old trap parts out. NOTE: If your trap is soldered copper or solvent welded plastic, cut the trap arm off squarely with a tubing cutter or hacksaw. Leave at least a 1¼ in. pipe stub-out to receive a mission coupling (See photo, lower left). Ream away any burrs or rim left by the cutting.

Attach the J-bend outlet on the P-trap to the downward-facing elbow of the trap arm. Slide the trap arm back into the trap adapter in the wall as far as it will go. Align the inlet of the J-bend under the outlet of the basket strainer or pop-up waste body above the trap. You may need to trim the trap arm. Smooth the cut edges.

With the trap inlet aligned under the basket strainer or pop-up waste outlet, measure the length your tailpiece would need to be. Cut and assemble the tailpiece, the end-outlet tee, and the tailpiece to match this length. Temporarily attach the continuous waste 90 to the tailpiece of the distant basin to determine where it will need to be cut to insert in the end outlet tee. Cut the continuous waste 90 and attach it temporarily to the end-outlet tee. Determine where the tailpiece on the distant basin needs to be cut to provide a slight slope to the continuous waste 90. Cut the tailpiece of this basin. Smooth and attach all the tubes with the slip nuts provided, using Teflon tape or pipe joint compound on tube threads and slip nuts, respectively. Remember that wedge-shaped washers point toward the joints.

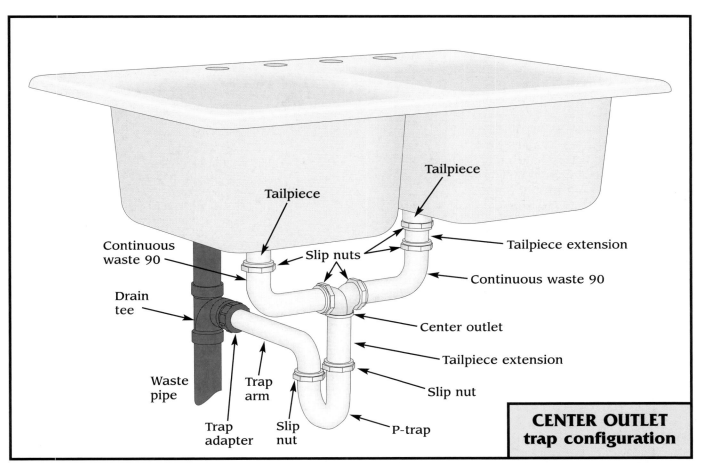

Tailpiece

Tailpiece

Continuous
waste 90

Slip nuts

Tailpiece extension

Drain
tee

Continuous waste 90

Center outlet

Tailpiece extension

Waste
pipe

Trap
arm

Slip nut

Trap
adapter

Slip
nut

P-trap

**CENTER OUTLET
trap configuration**

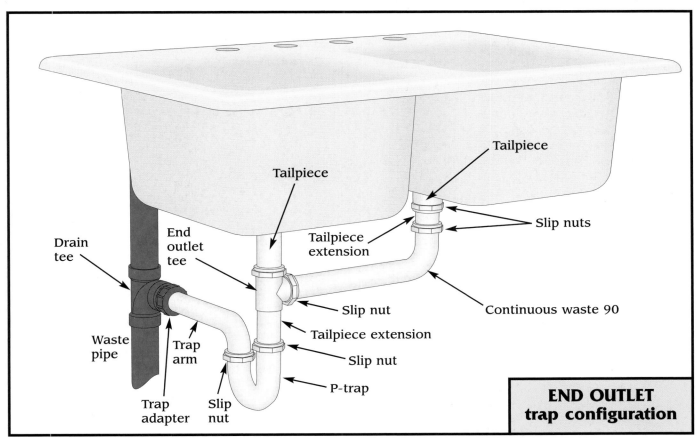

Tailpiece

Tailpiece

Drain
tee

End
outlet
tee

Tailpiece

Slip nuts

Tailpiece
extension

Slip nut

Continuous waste 90

Waste
pipe

Trap
arm

Tailpiece extension

Trap
adapter

Slip
nut

Slip nut

P-trap

**END OUTLET
trap configuration**

Installing a center outlet trap

Most sink traps for double-bowl basins are installed in a center outlet trap configuration (See top illustration, previous page). As with any sink trap installation, installing the parts correctly involves quite a bit of dry-testing fits, trimming and even a little trial and error. If you're installing a drain trap for a single-basin sink with only one drain, skip steps 2 and 3 in the following sequence and proceed as if the trap riser in step 5 is the drain tailpiece. If you're installing an end-outlet trap, see the instructions on page 98.

NOTE: This project, as shown, assumes you have pop-up waste assemblies with tailpieces on each lavatory or basket strainers with tailpieces on double kitchen sinks. It assumes you have checked with a local building inspector and determined: that the distance from the lavatory farthest from the drain stub-out to the stub-out is not great enough to require the installation of a second trap, and that you need (or don't need) baffles in your slip tee.

Before you being, draw your system and label all the parts you need. Make sure the waste components you buy are long enough. Each continuous waste 90 (they may be different lengths) will need to stretch from the tailpiece to the J-bend inlet. The trap arm, when fully inserted in the wall, should bring the J-bend out far enough to intersect with the slip tee. The outlet on the slip tee (or a slip joint extension tube) must insert at least an inch into the J-bend.

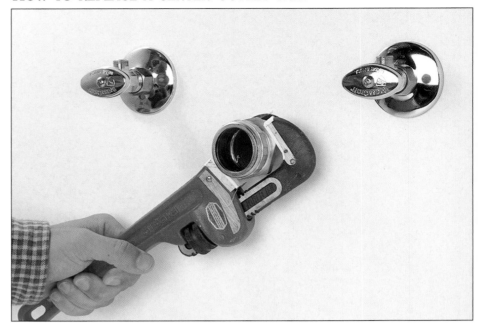

1 **Disconnect the old trap** assembly at the basket strainer or pop-up tailpiece and at the wall stub-out. Remove the old fittings—if necessary, cut the stub-out (very carefully) leaving at least 1¼ in. projecting out from the wall. Whenever possible, choose replacement parts that are the same material and size as the original. Attach a trap adapter to the drain stub-out in the wall. The adapter shown here has a threaded brass end that turns onto a threaded nipple from the wall. The other end has a slip nut for attaching to metal drain tubing. See "Transition fittings" on page 22 for more information on trap adapters to help you choose the correct fitting.

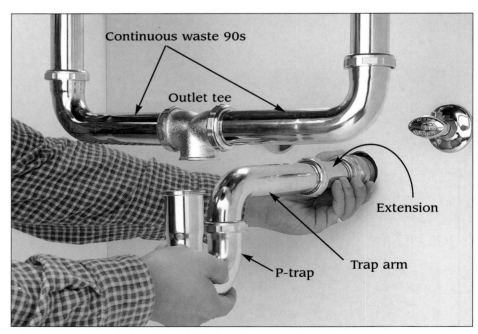

4 **Insert the ends of the trimmed 90's** into the center outlet tee so they are flush against the lips inside the tee openings. Then, reattach them to the tailpieces with slip nuts. Loosely attach the P-trap and trap arm to the trap adapter at the wall. The trap arm should be sized so the drain opening on the P-trap is aligned directly below the opening on the outlet tee. This will likely require you to either add an extension to the trap arm or trim the trap arm, or perhaps both. Some adjustment may be made by rotating the P-trap—it doesn't necessarily need to be parallel to the trap arm as shown. This step generally requires a bit of trial-and-error.

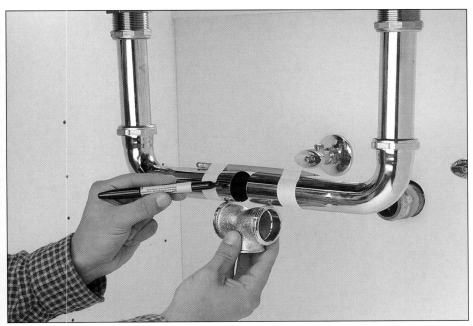

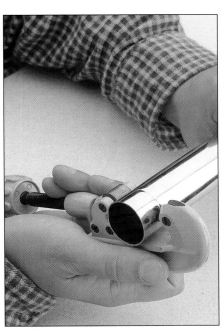

2 At the drain opening, attach continuous waste 90's (loosely) to the tailpieces with slip nuts. Tape a pencil or wood scrap so it spans the open ends of the 90s, aligning them. Hold the center outlet tee up to the gap between the ends of the 90's, making sure the tee is aligned with the trap adapter. Mark the ends of the 90's for trimming so they fit into the tee, seating against the lip inside the inlets on the tee. Remove the 90's, trim them as needed, and attach them loosely to the tee with slip nuts. Reattach the 90's to the tailpieces.

3 Remove the 90's and trim them as needed. We used a tubing cutter to cut the chromed brass tubing. You could also use a hack saw, but in either case be careful. The thin tubing is easily bent or creased from the cutting pressure.

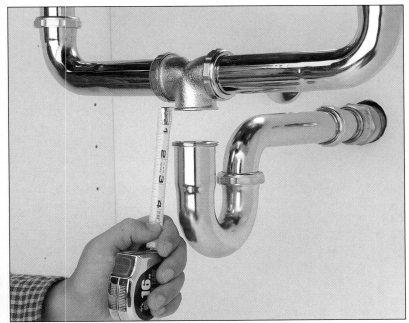

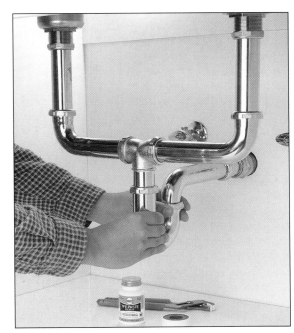

5 Once the P-trap assembly is trimmed, loosely attached to the adapter and aligned directly beneath the outlet tee, measure from the tee to the opening in the P-trap to determine how long to cut a trap riser to span the distance. Allow for the extension to penetrate at least ½ in. into each fitting. Allow for a slight downward slope of the continuous waste 90s. Cut the riser to fit and install it loosely with slip nuts at both the tee and the P-trap. Double-check to make sure all parts fit together without binding.

6 Remove the trap parts, then reattach them in sequence, applying Teflon tape to the male threaded members and pipe joint compound to the female threaded members. Hand tighten all slip nuts. Then, when everything is in good shape, use slide-jaw pliers (protect the jaws with masking tape) to tighten each connection firmly. Take care not to overtighten, however.

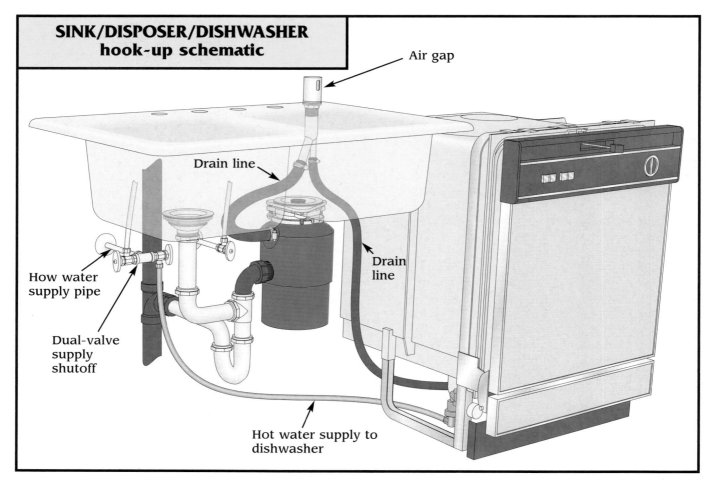

**SINK/DISPOSER/DISHWASHER
hook-up schematic**

Air gap

Drain line

Drain
line

How water
supply pipe

Dual-valve
supply
shutoff

Hot water supply to
dishwasher

Hooking up disposers & dishwashers

Note on wiring. Both disposers and dishwashers must meet local electrical codes when installed. In brief, a disposer requires a 15 or 20 amp, 115-volt circuit. If you may at some time install a dishwasher, you should install a duplex 20-amp receptacle—but check your codes first. Some require that each appliance have it's own circuit. Whether the appliance is plugged into a receptacle or hard-wired, both must have GFCI protection. A switch-operated disposer must have a switch wired into the circuit, at either a junction box or a receptacle. With a double receptacle, wire a switch only into the receptacle that will receive the disposer plug. The switch should be out of reach of children. Refer to the manufacturer's installation instructions for specific wiring information. If you are not comfort-

able working with wiring, leave this task to a professional.

Installing a dishwasher. In addition to a 15 or 20 amp direct line or GFCI three-pronged plug receptacle, a dishwasher requires a hot-water supply hookup and a drain line with an air gap. An air gap is a ventilated chamber installed on the deck of the sink that eliminates the risk of the dishwasher sucking dirty water out of the disposer or waste pipe and into the dishwasher. A ⅝-inch drain hose from the dishwasher drains into a larger, ⅞-in. drain hose at the air gap. The larger hose then drains into the disposer or the tailpiece of a basket strainer. The air gap can sit in the hole for a sprayer or in any free punch-out hole on the sink, or you may drill an extra hole in the sink for the air gap (See page 85). The air gap should be within the sink

deck, since a clog can cause water to come out the air gap. The hot water supply generally draws from the same supply branch as the hot water tap on the sink.

The opening for the new dishwasher should be about ½ in. wider than the dishwasher, which is typically 24 in. wide. The installation instructions may tell you where to cut the holes in the sink cabinet. Your goal is to allow the drain line to stretch from the nipple on the pump beneath the machine to the air gap on the back of the sink without kinking. Drill a pilot hole in the side of your cabinet and cut a 1½ in. hole in the cabinet using a drill and a hole saw when you are confident that your pilot hole is positioned right. Slide the dishwasher in to see if the hole will work for both drain and supply lines. Use a level at this time to adjust the front feet of the dishwasher and to determine if the rear feet will need to be raised or lowered. Pull the dishwasher out and adjust the rear feet as best you can. Pull the smaller drain hose and the supply tube through hole. While sliding the dishwasher back into place, feed the tubes through the intended opening in the dishwasher frame and out through the access panel opening in the front of the machine. If the dishwasher is still level, tighten the lock nuts on the adjusting screws of the feet. Make final adjustments to the back feet and tighten the lock nuts.

Attach the drain hose to the hose nipple on the pump beneath the dishwasher, positioning a hose clamp over the end of the hose. Tighten the clamp in place with a screwdriver or a hex nut driver. Wind Teflon tape clockwise onto both nipples of the brass elbow that will join your hot-water supply line to the solenoid valve of the dishwasher. Tighten the elbow in place on the solenoid, then tighten the braided stainless steel supply line to the elbow. Do not over-tighten this delicate connection. Leave the service panel off the dishwasher until you can ascertain that none of the connections are leaking when the supply line is on.

If you need to make new supply hookups, add a ⅜-in. stop extender tee to the hot water supply shutoff. This threads onto the ⅜ in. compression outlet of your hot water supply shutoff and provides an outlet for the dishwasher line as well as the sink. If your local code requires that the dishwasher have its own shutoff valve, replace the hot water angle stop with a dual-valve supply stop or put a tee on the supply stub out or nipple and add an individual supply shutoff. Apply Teflon tape clockwise to the threads of the supply shutoff and tighten it ½-turn past hand-tight.

Hold the head of the air gap in the hole to determine where to cut the narrower, dishwasher drain hose, which attaches to the nar-

rower leg of the air gap. Cut the dishwasher drain hose to length and attach it to the narrower leg of the air gap with a hose clamp. Attach the larger, uncut, air-gap drain hose to the wide, sloping leg of the air gap. Insert the air gap in the hole in the sink and secure it in place with the mounting hardware provided. If you will be attaching the air-gap drain hose to the tailpiece of a basket strainer, replace the tailpiece with a side-inlet tailpiece designed to receive waste from a dishwasher. Curve the air gap drain hose to meet the nipple on the disposer or basket-strainer tailpiece and mark to cut. The width of the hose should not be constricted anywhere, and the hose should slope continuously down from the air gap. Cut the hose and secure it to the drain nipple with a hose clamp. Important: If you are connecting a dishwasher to a disposer that has not previously drained a dishwasher, you will need to punch out an internal metal plug in the drain nipple with a screwdriver and hammer. Make sure you remove the punched-out plug. Make electrical hook-ups according to the manufacturer's instructions.

Attach the dishwasher to the underside of your counter by driving screws through the mounting tabs. Provide power to the dishwasher. Turn the water on. Look for leaks. During and after running the dishwasher, check for leaks with a flashlight. Check for leaks again after 24 hours. Tighten or replace leaking fittings.

Installing a disposer. Remove the mounting assembly from the new disposer: Insert a screwdriver into the right side of one of the mounting lugs on the lower ring of the mounting assembly to keep it from turning. With your free hand, turn the mounting assembly counterclockwise to remove it from the lower ring. If you will be using the new mounting assembly, disassemble it. Then, press a ¾ in. thick snake of plumbers putty around the

base of the flange. Press the flange into the drain hole with the manufacturers name right-side-up. Have a helper hold the flange in place from above. Slide the components of the mounting assembly on from below. Typically, you'll follow a fiber or cardboard washer with a steel pressure plate (flat side up), with a screw plate, and, lastly, a split-ring retainer to hold the plates up. Open the ring and slip it into a groove on the flange body.

Next, push the pressure plate up against the bottom of the sink. Turn the screws to widen the distance between the pressure plate and the screw plate. You'll want to alternate between screws as you tighten to achieve even pressure.

Attach the disposer. Your disposer will have a mounting ring on top that will insert and twist onto the screw plate of the mounting bracket. As you lift the disposer to the mounting bracket, align the mounting ears with the openings between the sloped mounting tracks of the screw plate. Twist the body of the disposer to seat the mounting ears on the sloped tracks. Make sure all three ears are resting on the screw plate. Insert a screwdriver into the left side of one of the rolled metal lugs in the mounting ring and twist the ring to the right. The mounting tabs will lock in place over the ridges.

Use a rubber gasket, metal flange, and screw to attach the new discharge tube. Rotate the disposer on the mounting ring until your discharge elbow faces the trap inlet. Attach the disposer discharge tube directly to your trap or to an end-outlet slip-joint tee with a slip washer and nut.

Attach the dishwasher hose to the dishwasher inlet nipple on the disposer with a hose clamp. Make the electrical hookups according to the manufacturer's instructions. Fill the sink half way with water, turn on the disposer and unplug to check for leaks from the mounting assembly and in the waste system.

Use a plunger as a first line of defense against clogs in your kitchen sink drain. Since most kitchen sinks have double bowls, you'll need to have a helper press the basket strainer plug down into the unclogged drain (the pressure from the plunging will dislodge it, breaking the pressure seal). Try a series of vigorous downward thrusts followed by an abrupt upward pull. Be patient. It may take a good 20 minutes of work to dislodge the clog. When you're done, flush the drain with hot water. Additional plunging at this time can help remove any remainders of the clog.

Unclogging sink drains

Bathrooms have lavatories while kitchens, bars, and laundries have sinks, but they all have drain openings, P-trap wastes, and P-trap arms that can become clogged (unless they are quite old and have S-traps). You'll follow slightly different steps for unclogging a lavatory than for a kitchen sink, however, since the lavatory usually has a pop up stopper and an overflow channel. The first "repair" attempt to remove a kitchen sink drain clog should be with a plunger.

When plunging fails, you'll need to remove the drain trap. Sink and lavatory P-traps that are physically accessible should be removable. If yours is not reachable, replace it. Loosen the large slip nut holding the trap arm in the wall with the largest wrench you have or can borrow. The nut may be hidden under an escutcheon at the wall. No nut? You may cut the trap arm near the wall with a hack saw or jig saw, leaving an inch or more of pipe (not fitting) stub to attach a mission coupling to. The tailpiece of your lavatory unthreads and is replaceable, if this is necessary. Replace the old trap with a slip nut trap.

HOW TO UNCLOG A KITCHEN SINK TRAP

1 **Put a bucket under the trap** and loosen the slip nuts holding the trap on. Clean out the trap, using a wire brush to remove any build up. Remove material from the trap arm as well.

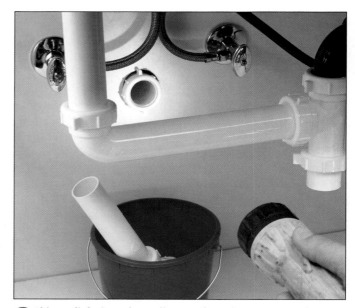

2 **Shine a light into the wall** to see if standing water is in the sanitary tee, indicating that the waste pipe in the wall is clogged. For a clogged drainpipe call a drain cleaning service or try to clear it with a hand auger (See page 52).

How to Unclog a Lavatory Trap

Before taking drastic steps, try plunging. To plunge the trap, first stuff a small wet sponge in a plastic bag into the overflow opening. Remove the stopper but leave in or replace the pivot ball. Put enough water in the sink to cover the plunger, then pump the plunger up and down over the drain, pulling the plunger up abruptly after each six or seven plunges. If fifteen sets of plunges don't clear your drain, you will have to remove the trap. Replace the pop-up stopper, and remove the rag from the overflow. Adjust the lift strap so the pop-up stopper works properly.

1 **Remove the pop-up drain stopper.** You may have to turn a pop-up stopper counterclockwise to remove it, or you may have to undo the retaining nut in the back of the pop-up stopper body and pull out the stopper pivot rod, if this retains the stopper post. Clean hair and soap buildup from the stopper post with a wire brush or other appropriate tool. If your trap is not removable and you will need to plunge, leave the stopper out, but replace the pivot rod, washer, and retaining nut to restore the drain seal for the plunger.

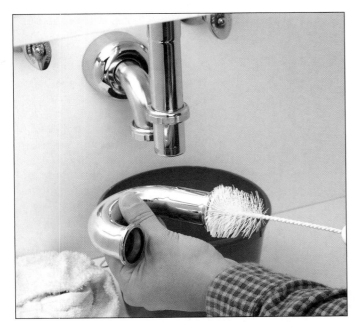

2 **Put a bucket under the trap** and loosen the slip nuts holding the trap on. Clean out the trap, using a wire brush to remove any build up. Remove material from the trap arm as well. Use a coat hanger with a hook bent into it, or mechanical fingers. Remove the trap arm if this is helpful.

3 **Look into the drain** fitting in the wall with a flashlight. Standing water indicates a plugged drainpipe. Call a drain service or try using a hand auger. After successfully clearing the clog, clean threaded parts of the trap with a wire brush if needed, and apply pipe joint compound to the female threads and Teflon tape (wrapped clockwise in the direction of the threads) on the male threads before putting the trap back on. If your trap is plastic, you don't need the pipe joint compound.

Unclogging disposer drains

Cleaning a disposer. Running enough cold water when using a disposer should keep it clean, but at times the top of the grinding chamber will become encrusted with food particles and grease. To clean it, turn off the disposer and disconnect the power supply. Reach into the disposer and clean the underside of the splash baffle and the upper reaches of the grinding chamber with a scouring pad. Then place the stopper in the opening, put ¼ cup of baking soda in the sink, and fill the sink half way with warm water. Turn on the disposer and remove the plug to flush out the disposer and to neutralize odors.

Restarting an overheated disposer. Disposers turn off automatically when they overheat. Wait two to five minutes to let the machine cool, then hit the red RESET button. If this doesn't work, check your house service panel to see if a circuit breaker tripped or a

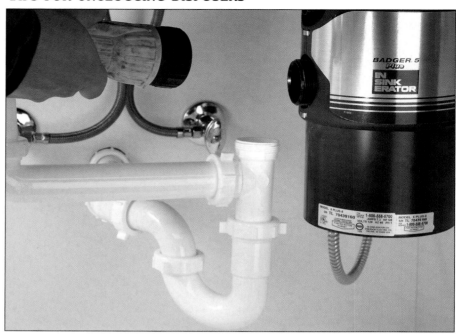

If the blockage is in the disposer, first turn the fuse or switch off at the circuit breaker and unplug the disposer. Remove the discharge arm from the disposer and see if you can see the grinding plate with a flashlight.

fuse blew. Otherwise, the machine could be jammed.

Removing stuck objects and freeing a jam. Turn off the power to the disposer and remove the splashguard. Try to locate the stuck object with a flashlight. If you can't remove an object with tongs or your hand, or if no object is visible, attempt to rotate the grinding plate one full rotation. Try rotating the plate in either direction by levering a broomstick against the deck of the sink and the impellers on the circumference of the plate.

Note: Some disposers come with special wrenches that fit in a hole on the bottom of the disposer to rotate the grinding plate. It is also possible to buy a tool especially for turning the grinder.

Bad News: If your disposer is choking on banana peels, string, or some other fibrous waste, you'll probably need to have it taken apart by an appliance repairperson. It might be cheaper to replace the disposer.

Standing water in the sink basin containing the disposer is a sure sign that drain clogs can be attributed to the disposer drain line (presuming that the other side of the sink drains freely).

Use needle nose pliers to remove material plugging the disposer.

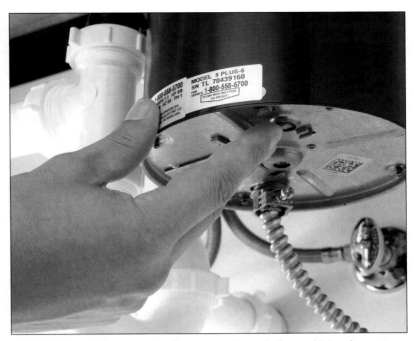

If the blockage is because the disposer won't work, first try hitting the reset button on the disposer.

Dishwasher drain problems

Dishwasher drainage problems include clogs, leaks, and clogs that cause leaks. See if the problems described below sound familiar.

Leaks from the door can sometimes indicate a clogged drain line, causing too much water to remain in the machine. Leaks from the air gap indicate a clog in the air gap or on the outlet-side hose of the air gap. Leaks beneath the sink or dishwasher indicate a cracked hose or a loose hose clamp.

A clogged or kinked line from the air gap to the disposer or basket strainer tailpiece may cause water to leak from the vent slots in the air gap. Clogs may leave standing water in the dishwasher. If the air gap doesn't leak, but the dishwasher doesn't drain well, remove and clear the hose between the dishwasher and the air gap or see if the drain screen is clogged. Or you may need to replace a drain line solenoid that's designed to protect against backflow.

Drain line leaks usually occur when hose clamps work loose or break. Tighten or replace hose clamps at the pump under the machine, at the disposer or tailpiece, or at the air gap.

Try rotating the grinding wheel of the disposer with a broomstick before turning it on again.

Tubs & Showers

Installing bathtubs and showers has gotten much easier over the years. It wasn't too long ago that any shower required a custom-poured mortar base with metal reinforcement. Drains were fashioned from oakum and other hard-to-handle materials. Today, the availability of do-it-yourselfer friendly kits has greatly simplified the process. And new advances in plastics and polymers have brought us newer, lightweight alternatives to traditional cast iron and porcelain. Bathtubs, for example, are now made from polymers that have many of the same properties of much heavier materials, as well as plastic, fiberglass, and lightweight pressed steel.

Shower and tub surrounds are almost always installed as kits these days, compared to the clunky, site-built efforts that always seemed to leak or lose their surfacing. Along with these advances, some new accessories have been developed, including prefabricated glass doors that are a breeze to mount.

But one outcome of the flood of new products in this area is that every product you buy seems to have slightly different installation methods. Standardization has become a thing of the past. In the sequences contained in this section, we show you some examples of how these new materials and kits are installed. But as likely as not, the unit you buy won't be installed in exactly the same way. Nevertheless, we've tried to select fairly representative products to work with and to present them in a way that emphasizes the general ideas over the specific details. When combined with the manufacturer's instructions that come with your product, you should have all the information you need to do the job right.

Repairing tubs and showers can get a little tricky, mostly for reasons of access. While plumbing codes require that all plumbing connections for these fixtures be accessible, often through a removable wall panel, tubs and showers in older homes may not have any access. Or if they do, the way in may be through a ceiling below without enough access to be of any use. Even if your tub or shower plumbing appears to be working just fine, it's a good idea to inspect the access panel. If there is none, add one. If the existing opening is too small, enlarge it. Not only will you appreciate this if the plumbing develops a problem in the future, but getting into the opening will let you inspect the plumbing itself. Otherwise, small leaks in hidden plumbing can go undetected for years. You won't find out about them until a lot of damage has occurred.

CROSS-SECTION OF A SHOWER AND BATH HOOK-UP

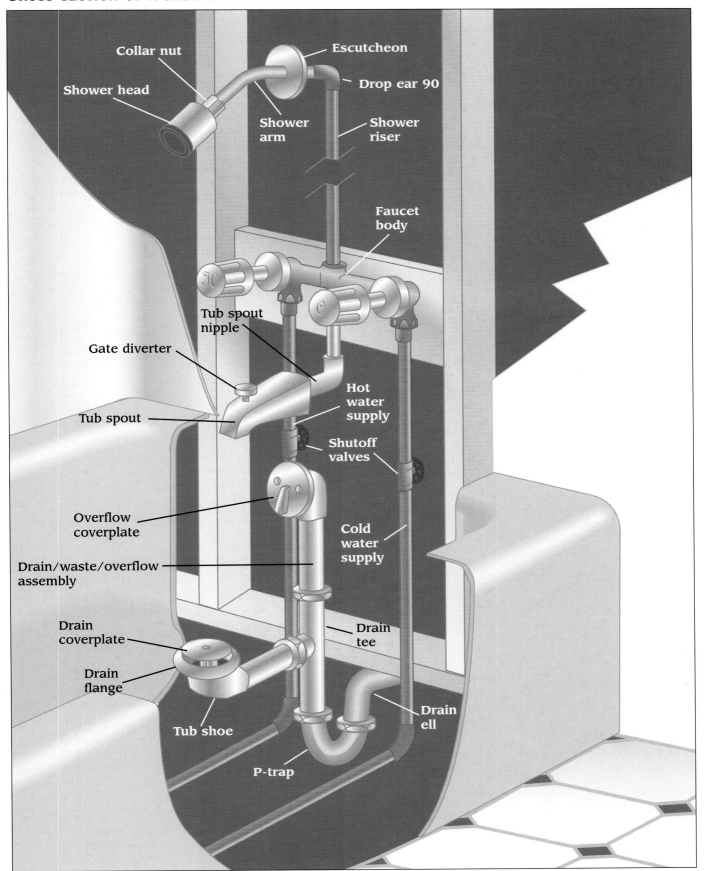

Collar nut

Shower head

Escutcheon

Drop ear 90

Shower arm

Shower riser

Faucet body

Tub spout nipple

Gate diverter

Tub spout

Hot water supply

Shutoff valves

Overflow coverplate

Cold water supply

Drain/waste/overflow assembly

Drain coverplate

Drain tee

Drain flange

Tub shoe

Drain ell

P-trap

Diverter valves, located in or at the connection to the tub spout, close off the flow of water to the tub spout and direct it up through the riser to the shower head. The diverter, faucet and spout work together and are typically sold as a complete package.

Diverters & faucets

Shower and tub faucets are larger versions of sink and lavatory faucets, with the exception that combination tub/showers have manual diverter valves to switch water flow between the tub spout and the showerhead. The diverter valve may be controlled with a handle in the wall or with a lever on the spout. Three common styles include: one handle hot/cold control with a diverter valve on the spout; two-handle hot and cold control with the diverter on the spout; and three-handle hot, cold, and wall-mounted diverter control.

Wall mounted diverters may use compression valves, like two handle faucets. If this valve fails to switch water flow completely to the shower or completely to the tub, fix it. TIP: To keep handles from corroding to stems, put valve stem grease on the flutes of the stems and in the slot on the handle, then put Loc-tite on the handle screw to keep the handle from loosening.

DIVERTER TYPES

One-handle hot/cold control with a diverter valve on the spout.

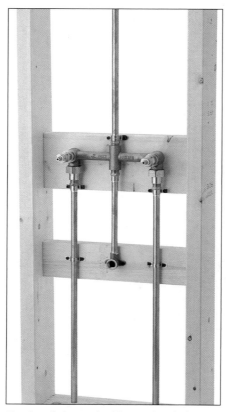

Two-handle hot and cold control with the diverter valve on the spout.

Wall-mounted diverter control.

HOW TO FIX A WALL-MOUNTED DIVERTER

1 Shut off hot and cold water supply. Pry off the protective cap on the diverter valve handle then remove the screw behind the cap on the end of the diverter handle to remove the handle. You may need a handle puller if the handle has become corroded in place (See page 66).

2 Unscrew or pry off the escutcheon at the base of the valve stem. This will give you access to the bonnet nut that secures the diverter valve. Remove the bonnet nut with an adjustable wrench then remove the diverter assembly with a deep-set socket wrench.

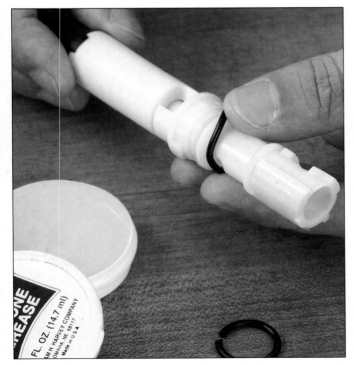

3 If the diverter uses compression washers and rings, replace these. If the diverter valve uses a washerless cartridge, like the one shown above, you need to purchase a replacement parts kit for your diverter model. Or, simply replace the entire cartridge.

4 If you are replacing neoprene parts, coat replacements with heat-proof grease. Replace the diverter. Press the escutcheon into a ring of plumbers putty before securing it to form a seal at the wall.

1 Spout-mounted diverters can't be fixed. You must replace the entire spout. Look for a slot under the handle to access an Allen screw. Remove the screw and slide off the spout.
NOTE: If all you have to attach your spout to is a smooth copper stub-out, you will need to get a replacement spout that also uses a setscrew. If you have a threaded brass or steel nipple for a stub out, buy a solid brass, female-threaded replacement spout.

2 If there are no setscrews, the spout will unscrew from a brass or steel nipple. Use a large screwdriver, a hardwood dowel, or grip the spout with pipe wrench to unscrew the spout. Find a comparably sized replacement spout made of chrome-plated brass. Tub spouts sometimes get used as safety handles and need to be strong.

3 Clean the threads of the nipple and wrap them with about four layers of Teflon tape in a clockwise direction, working from the tip back (top photo). Pack the cavity in the back of the new spout with plumbers putty to form a seal with the tub surround. Apply pipe-joint compound to the female threads of the spout (bottom photo).

4 Screw new spout in place using a hardwood dowel or broomstick to tighten it on. Depending on the position of the threads, you may be able to tighten it with your hands alone.

HOW TO FIX A LEAKY SINGLE-HANDLE TUB/SHOWER FAUCET

NOTE: Single-handle faucets for tubs are like single-handle faucets for sinks: you just remove them and turn off the water a little differently. Find your single-handle valve type under "Faucet types" on pages 64 to 65 to reference the instructions for your valve type.

Testing a tub & shower for pressurized leaks

To tell if your tub has a pressure leak (from the pressurized hot and cold water system) or a gravity leak (from the waste water drain system) you might try the following test.

1 Hook a hose up to the showerhead arm using a hose adapter. You might have to replace the shower arm with a galvanized nipple to attach the adapter. Run the other end of the hose out the window.

2 Turn on the hot and cold water, diverted to the shower, and let run. If the leak shows up, then you know you have a pressurized water leak since the drain system has been bypassed. Experiment with running only cold or only hot to hone in on the problem.

3 If it happens with hot or cold, try wrapping Teflon tape clockwise on the shower arm where it threads into the fitting in the shower wall. If it could be a valve problem, remove the handles and escutcheons from the diverter or the hot or cold handle or a single handle to see if water is escaping through the packing or a worn O-ring and leaking down inside your wall.

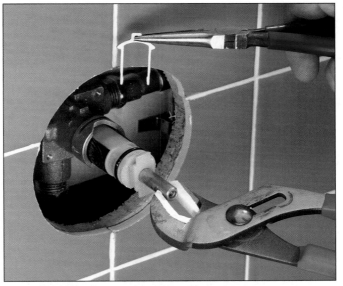

1 Remove the handle and the escutcheon with a screwdriver. You may have to lift off a cap to remove the screw holding the handle on. You may be able to turn off the hot and cold water to the faucet using screwdriver operated shut-off valves right behind the escutcheon. Otherwise, turn off the hot and cold water in the basement or through an access panel.

2 Remove any nuts, rings, clips, or caps, holding the cartridge or ball in place. Refer to pages 64 to 75 to reference instructions on replacing parts for your valve style.

3 Clean any sediment buildup found and flush clear with water before coating new parts with heatproof grease and installing. Apply plumbers putty to seal behind the escutcheon.

How to fix a leaky two-handle faucet

1 After turning off the water through an access panel or in the basement, remove the screw behind the cap on the hot or cold-water handle, to allow you to remove the handle. You may need a handle puller if the handle has become corroded in place. Unscrew or pry off the escutcheon at the base of the valve stem. Unscrew the escutcheon with tape-padded, slide-jaw pliers.

2 To access the bonnet nut, it may be necessary to chip out mortar with a hammer or mallet and a cold chisel. Carefully chip away only enough material to allow a socket to fit into the opening and be secured around the bonnet nut.

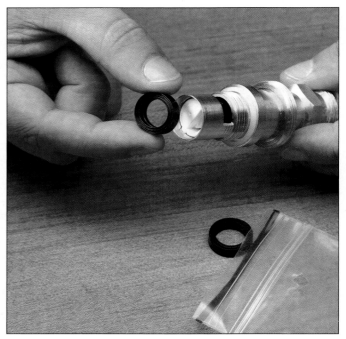

3 Use a deep socket (the one shown here is a non-ratcheting plumber's socket specially designed to fit over valve stems or cartridges and access bonnet nuts) to unscrew the bonnet nut and stem or cartridge. TIP: If the bonnet nut that secures the valve stem or cartridge is loose, this could have permitted a behind-the-escutcheon leak. Try tightening the bonnet nut and turning the water on to see if this solves the problem.

4 A leaking cartridge can be repaired by replacing O-rings and/or packing material and packing washers, as is being done above. But most plumbing experts will suggest that you simply discard the old cartridge and replace it with a new one. If you decide to repair your cartridge, keep all components in descending order when disassembling it, then record this order on paper.

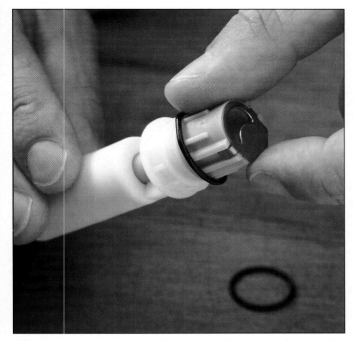

5 When repairing a cartridge, be sure to replace O-rings that seal the cartridge in the faucet body, as well as any interior springs and seals. Remember to coat all new neoprene parts with heatproof grease.

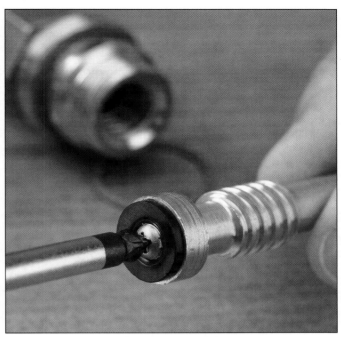

6 Bring compression stems and parts with you to a plumbing supply store. Purchase all new neoprene parts and a new stem screw. Apply heatproof grease or petroleum jelly to neoprene parts before reassembly.

7 Reinstall the valve/cartridge, then tighten the bonnet nut with a socket to secure it.

8 Reattach the escutcheon and handles. Use a generous amount of plumbers putty to keep water from getting behind the escutcheon.

Clearing & fixing tub drains

Tub and tub/shower drains come in three styles. In the simplest formation, you plug the tub drain with a *rubber stopper* that fits in the drain like a cork. A *plunger-style* (lift-bucket) drain uses a tube-shaped plunger (also called a bucket), which slides in front of the tee that joins the drain shoe to the vertical pipe in the drain assembly. This kind of drain stopper is hidden except for the trip lever. A *pop-up style* tub stopper rides up and down on a rocker arm in the shoe, corking the drain at the tub outlet like an old-fashioned stopper. Both plunger and pop-up stoppers are hooked to a trip lever over the overflow drain opening with a linkage through the overflow drainpipe.

Tub drains most often become clogged within the drain assembly. Typically, the drain linkage, the plunger, the drain strainer or the pop-up stopper becomes tangled with hair and soap. Remove these parts and clean. Alternatively, the drain shoe and trap may be plunged, but first you must plug the overflow opening to create suction (See photo below). Sometimes clogs in and beyond the trap can be cleaned out by augering through the overflow or by augering the branch drain through a cleanout.

Plunger and pop-up stopper drains can also break or come out of adjustment. If the stopper doesn't seat well or open far enough, cleaning, adjusting and lubricating the linkage may be enough to get it working again.

TIP: When using a plunger to try and clear a bathtub drain, plug the DWO opening in the tub before plunging. A damp sponge sealed in a plastic bag works well as a plug for the opening. The plug seals the drain system (up to the trap, anyway) so suction can be created.

Unclogging tub & shower drains

Shower Drains

Like tubs, shower drains become clogged with hair and soap. Unlike tubs, you don't have to deal with an overflow or a pop-up stopper. If your shower drain is clogged, follow these steps:

1 Start by removing the strainer that covers the drain opening—some are screwed in place and some are held in with pressure-fit tabs. Completely clean hair and soap from the strainer and set aside. If the buildup in the strainer is significant, test the drain to see how well it works once the clog is removed. If the drainage is still slow, proceed to step 2.

2 Shine a flashlight into the drain opening and dislodge any blockages you see with a tight wire hook or mechanical fingers (See photo 2).

Inspect the drain opening with a flashlight. Use wire or mechanical fingers to remove any clogs you spot.

3 If drain is still blocked, run water into the shower pan, deep enough to cover the cup of a plunger. Pump vigorously with the plunger, pulling back abruptly after each six or seven pumps. Repeat as necessary. Run hot water into the drain and continue plunging if drainage is still slow.

4 As a last resort, try clearing the blockage with a "snake" type hand auger.

Use a plunger to try and force out difficult clogs. Make sure to fill the shower pan with enough water to cover the plunger cup first.

Plunger-type drains

A plunger-type (lift-bucket) drain uses a tube-shaped plunger (also called a bucket), which slides in front of the tee that joins the drain shoe to the vertical pipe in the drain assembly. The mechanism is hidden except for the trip lever. If your plunger-type drain is clogged, follow these steps:

1 Remove the strainer from the drain. Completely clean hair and soap from the strainer and set aside.

2 Shine a flashlight into the tub shoe area and dislodge any blockages you see with a tight wire hook or mechanical fingers. Test the drain. If still clogged, proceed to step 3.

3 Use a screwdriver to remove the cover plate of the trip lever. Pull the plunger from the overflow opening (See Photo 3). Completely clean hair, soap, and mineral buildup from the plunger and linkage. Replace damaged parts. Coat parts, including the plunger, with heatproof grease

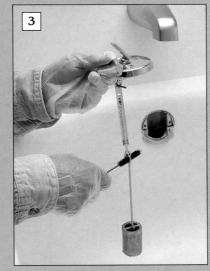

4 If the drain is still plugged, try using a plunger to remove the blockage. Stuff a wet sponge in a plastic bag into the overflow (See photo, previous page). With water in the tub to create suction, pump vigorously on the drain with the plunger. Plunge in series of six or seven pumps, ending each series with an abrupt upward pull. Repeat as necessary.

Remove the plunger mechanism from the overflow opening and inspect it for clogs or damage..

5 As a last resort, try clearing the blockage with a "snake" type hand auger.

6 Replace the plunger assembly, coverplate, and strainer. Flush the system with hot water. Additional plunging during the hot water flush can help remove pieces of the clog.

Pop-up drains

If your tub has a pop-up drain cover (the most common type of plug these days), remove clogs by following these steps:

1 Lift the trip lever to raise the stopper so you can pull the stopper and rocker arm from the drain hole in the tub. Clean the stopper and rocker arm with a small wire brush and vinegar (See photo 1). Replace a worn O-ring on the stopper. Coat moving, rubbing parts with heatproof grease. Test the drain. If clog persists, proceed to step 2.

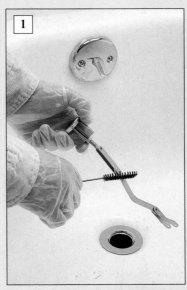

Remove the plunger mechanism from the overflow opening and inspect it for clogs or damage.

2 Remove the coverplate screws and pull the trip lever and its linkage from the overflow opening. Replace any damaged parts or, if wear is extensive, replace the entire pop-up assembly. Clean all interior parts and coat with heatproof grease.

3 If the drain is still plugged, use a plunger or a hand auger to remove the blockage. See steps 4 and 5, "Plunger-type drains".

4. Replace the rocker arm and stopper and the trip lever and linkage, adjusting as needed to create a tight seal (See Photo 4).

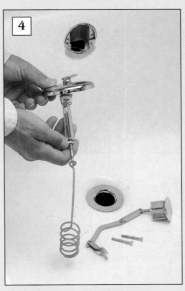

Replace the pop-up drain assembly after the clog is removed. Test the linkage to make sure the drain plug seals tightly. Adjust the rocker arm as needed.

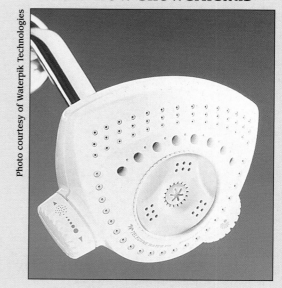

Low-flow showerheads

New showerheads are required by law to be low-flow. This will save the average household about 12,000 gallons of water a year. Lower quality heads or flow-restricting washers tend to atomize the water into an ineffectual mist. High-quality low-flow showerheads keep water droplets from becoming too small, delivering a shower that feels voluminous. Water massage shower heads pulsate the water flow—an action that saves water while producing a forceful shower. Showerhead shutoff valves reduce flow to a trickle with the flick of a lever, saving water while the bather lathers. Shutoffs can be purchased separately and are sometimes built into better showerheads.

Showerheads

Whether you start with a bathtub, a shower or an existing tub & shower you can create the perfect shower for your family without a lot of work and expense. Showerheads and shower arms unscrew and are easily replaced, and spout-mounted shower adapters are available for converting simple tubs into tub/shower combinations without the inconvenience of rubber hoses and end-of-spout connectors.

Hand-held showerheads that include a hanger for fixed operation may be installed on either a tub spout with a diverter valve or on a shower arm. These need to be equipped with an integral or added-on vacuum breaker to keep a submerged shower head from sucking dirty water out of the tub, should the municipal wastewater system back up due to a broken pipe.

Periodically, shower heads need cleaning, tightening, or repair. For leaks at the head or shower arm, replace O-rings, or remove the head or arm and rewrap threads with Teflon tape. Diverter valves to the shower may also need cleaning or repair, if water leaks from the tub spout while the shower is operating.

For the chlorine sensitive, there are now shower filters that remove chlorine or convert it to harmless, odorless chloride compounds. Note that shower filters do not remove dangerous organic chemicals and chemicals resulting from improper water chlorination practices. But if you just want to get rid of chlorine from shower water, look for non-carbon filters with replaceable cartridges and specific claims about chlorine reduction (See pages 155 to 157).

HOW TO CLEAN A SHOWERHEAD

1 Wrap tape around the jaws of your slip jaw pliers and remove the swivel-ball fitting holding the head to the arm. Unscrew the collar nut holding the swivel-ball to the head.

2 Clean outlet and inlet holes with a thin wire or soak the head in vinegar overnight to dissolve mineral deposits. Flush the head with water.

3 Replace the O-ring if the head leaked or didn't hold its position. Coat the new O-ring with heat proof grease. Hand tighten the swivel fitting to the head with the collar nut.

4 Apply Teflon tape clockwise to the shower-arm threads, then tighten the retainer or ball fitting half a turn past hand tight with your pliers. Tighten further if it leaks.

Convert a single-purpose tub to a combination tub shower by adding a hand-held shower head with a flexible tube.

Adding a hand-held shower adapter

You can convert a tub to a tub-and-shower with a flexible-hose shower adapter. Purchase a quality shower head or shower massager. The shower head can be mounted for shower traditionalists or it can be hand-held for greater control and impact. A spout with a diverter valve allows you to switch easily from tub spout to shower.

Hand-held showers use water more efficiently than fixed shower heads. Flexible showerhead tubes with mountable heads can be added to a regular shower arm as well as to a shower-adapter tub spout.

How to add a hand-held showerhead to a tub with surround

1 Remove the old spout. Look for a slot under the handle to access an Allen screw. Remove the screw and slide off the spout. If there are no setscrews, the spout will unscrew from a brass or steel nipple. Use a pipe wrench or insert the solid shaft of a tool or a hardwood dowel into the spout to unthread it.

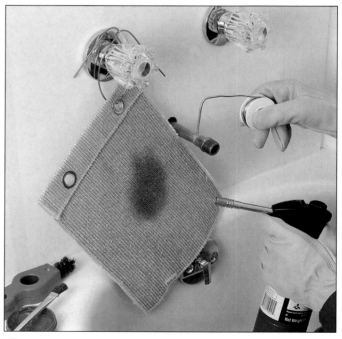

2 To supply water to the hand-held showerhead, you'll need a new spout with a gate diverter and an adapter hose outlet. Look for a spout that fits the existing nipple. If the old nipple is unthreaded copper, however, it's a good idea to cut the stub and solder a threaded adapter onto the end. That way, if you need to remove the spout for repairs or replacement in the future, you can simply unscrew it.

TIP: If you need to replace the nipple, buy ½ in. brass nipples (never use galvanized) in a range of lengths and return the ones that don't fit. After removing the old nipple with a pipe wrench, dry-fit nipples until you find one that allows the spout to be pressed firmly to the wall when tightened fully onto the threads. Wrap the threads of both ends of the new nipple, clockwise, with Teflon tape. Apply pipe joint compound to the wall-side of the nipple for added insurance against leaks. Thread the nipple into a spout with a shower adapter outlet, pack the spout with plumbers putty, and tighten the spout to the wall.

3 Attach a vacuum breaker and the adapter hose to the spout adapter-hose outlet. Wrap the male threads of the adapter hose outlet and the vacuum breaker clockwise with Teflon tape. Apply pipe joint compound to the female threads of the hose and the vacuum breaker and tighten the components together with an adjustable wrench.

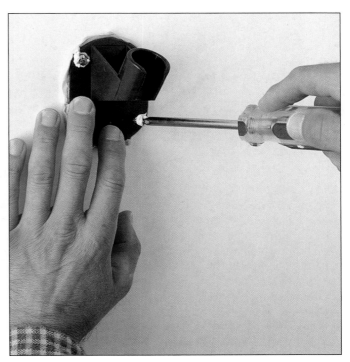

4 Position the showerhead at a comfortable height (65 to 78 in. is typical), then mark where the showerhead hanger needs to be. Leave adequate slack in the hose to lift the showerhead from its hanger. Mark drilling points for the screw holes in the hanger on the surface of the tub surround. Drill pilot holes for the screws. If the hanger is not located at a framing member, you'll need to use toggles or plastic screw anchors to secure the screws. If the surround is made of ceramic tile, rap a nail very lightly at the drilling point to make a starter hole (inset photo, top) then drill the hole with a masonry bit (inset photo, bottom).

5 Insert screws into the holes in the hanger and dab some silicone caulk onto the screw tips to seal around the hole. Drive the screws then caulk around the edges of the hanger. If your tub is built into an alcove, the easiest way to provide a shower curtain is with a telescoping, pressure-fit shower curtain rod. Otherwise, hang tracks from the ceiling to support the shower curtain.

Today's popular pressed steel or fiberglass tubs are manufactured to be installed in alcoves with manufactured tub surround panels. Installation is very DIY-friendly. Shower stall kits (inset photo) come in many shapes and styles. See how the one above is installed on pages 130 to 133.

Installing tubs & showers

Here we guide you through steps involved in replacing a bathtub, shower stall or combination tub-and-shower. These are not easy jobs and they require carpentry skills and tools. You will also need a helper—from removing the old tub through attaching the waste and positioning the new tub. Potentially high levels of demolition (and therefore, reconstruction) are involved. You will need to remove at least some of the old tub surround, possibly floor tiles, and possibly sections of wall and ceiling from rooms adjacent to or below the bathroom. You may need to cut a tub-sized hole in the wall

to remove the old tub and bring in the new unit.

In the project featured here, we replace an alcove tub with an alcove tub. The same basic steps are involved for replacing a tub with two or three open sides. Changing from an open tub to an alcove tub is easy enough if you know how to construct a partition wall. Alcove tubs are popular because the shower curtain can be replaced with easily cleaned, translucent sliding doors.

You'll also see how to replace tub and shower plumbing. You may replace the tub without replacing the hot- and cold-water plumbing.

However, if you want to convert a tub to a tub with a built-in shower, you must open the wall to replace the faucet and add piping to the showerhead. These steps are included here.

Code Considerations

• Make sure the position of your tub meets with local code requirements concerning minimum allowed distances of the tub from other fixtures in the bathroom.

• If the tub trap joints are made with slip nuts, or if they include a ground joint union, they must be accessible (such as through a removable ceiling or wall panel.) Non-accessible traps must use cemented, soldered or no-hub joints.

As a general rule, install a new

Before starting tub removal

• Close the hot and cold shutoff valves to the old tub. These valves may be found behind an access panel in the back of your "wet" wall (the wall containing the pipes) or in the basement or crawlspace below the tub. If no shutoffs exist, shut down the house water, turn off the hot water at the water heater, drain the pipes, and install shut-off valves on the supply lines. Preferably, position these stops behind an access panel in the wet wall.

• Remove the faucet handles and escutcheons, and the shower arm and escutcheon if a shower exists (the shower arm unscrews). Remove the old tub spout.

TIP: Making an access panel

To make an access panel, cut or break into the stud bay containing the pipes through the other side of the wall from the tub. Locate the studs framing this bay and cut out the wallboard evenly on the center of the studs. Extend the rectangle from the floor to a height of two feet. When the tub has been removed, frame the top of the opening with a 1 × 4 turned flat to the wall, then trim the opening with molding. Cover the opening with a removable, finish-grade plywood panel.

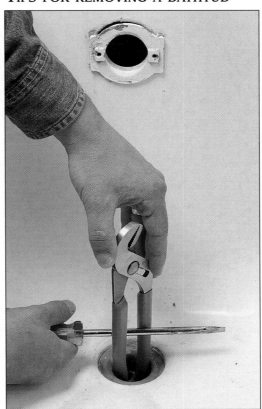

Remove the drain assembly by first unscrewing the overflow coverplate and removing any bracket or nut on the overflow that secures the overflow pipe to the tub. Use a strainer wrench or pliers handles and a screwdriver to twist off the drain flange. With a pop-up waste, you may need to get a special kind of strainer wrench that has slots to fit over lugs in the inside wall of the waste.

faucet and the piping for a new shower before installing the tub. This will involve some demolition and the installation of horizontal blocking for the new hardware. Here we use a two-handle valve with a wall-mounted diverter handle (See page 110), but you may use any diverter set-up you choose. Whichever valve style you purchase, avoid "fishing at the bottom of the barrel," since defective or worn out shower valves are not easily replaced. Buy quality fittings as well.

If you've had problems with banging pipes when turning off the old tub or shower, consider installing commercial or home-made air chambers on your tub and shower supply pipes (See page 54). These can tee off the supply pipes just above the shut-off valves. Use copper stubs and 90° elbows to extend the air chambers past and above the level of the valve body.

Conservation Tip: Much water is wasted while lathering in the shower. Showerhead shutoffs (installed above the shower head) reduce water to a trickle with the switch of a lever. The trickling water holds the temperature but drastically reduces water consumption in the shower while bathers lather.

Cut back or remove the wall covering in the work area. If the new unit is exactly the same size as the old one, you can use the existing wall stud structure—only remove the wallcovering up to a point about one foot above the tub flange. If the new tub is smaller (as in the project here), remove wallcoverings in the entire area so you can tack furring onto the wall studs. If you're installing a new faucet and shower plumbing, you'll need to remove the wallcovering from the wet wall at least as high as the showerhead.

TIPS FOR REMOVING A BATHTUB (CONTINUED)

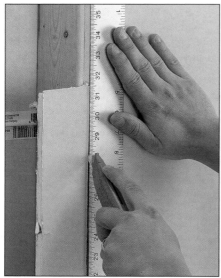

Be neat when removing wallcoverings. Locate the next wall stud outside the tub area on each side and score vertical cutting lines centered on the wall studs, then remove wallboard neatly up to the lines. This creates a nailing surface for patching in with new wallboard—be sure to use water-resistant, vinyl-covered wallboard (generally called "greenboard") for the new wall surface.

It may be necessary to rock the tub with a crow bar to break it free from a mortar bed. Pry the tub away from the wall with a wrecking bar and remove it, leaving the waste and overflow attached to the trap. Take care to protect your floor with boards. Slide the tub out of the alcove on soaped 1 × 4s if it's made of heavy cast iron. Cut caulk seals before removal (inset photo).

HOW TO INSTALL A NEW BATHTUB & SHOWER

1 Close the shutoff valve, remove the old tub and clear the wet wall of any wall covering from the level of the tub to the ceiling. Remove nails and any other hardware from the studs. Remove old faucet, pipes and hardware by cutting the supply tubes between the shutoff and the valve assembly, leaving plenty of exposed tubing to work with. Unscrew union nuts or release soldered pipes with a torch where hot and cold supply pipes enter the faucet body. If your stud bay for the plumbing is not wide enough to accommodate the new tub unit, move a stud or studs as needed.

2 Decide where the faucet body, spout, and shower arm fitting need to be and mark their positions on your studs. If you're replacing a regular tub with a tub/shower combination, you'll probably want to raise the position of the faucet body higher off the floor (32 to 36 in. is a typical height for shower/bath faucet handles). The tub spout should be about 4 in. above the rim of the tub, and the shower arm fitting typically ranges from 65 to 78 in. off the floor. This height will depend on the height of the people using the shower, and how far your shower arm and showerhead descend from the shower-arm fitting. Remember that the tub itself adds some height to the bather.

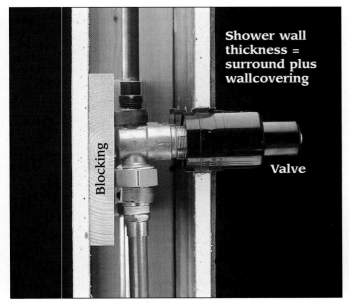

Shower wall
thickness =
surround plus
wallcovering

Blocking

Valve

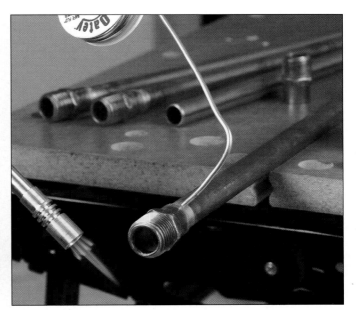

3 Determine where the blocking needs to be. The valve body, shower-arm fitting and tub-spout fitting each need a backing block. In a 2 × 4 wall, you will need to use 1× dimension lumber to provide clearance for the valve. In a 2 × 6 wet wall you can use 2× dimension lumber. The position of the blocking for the valve is the most critical. To locate it correctly, first calculate the finished thickness of the tub-surround by adding the thickness of all the layers (such as cement board and tile). Subtract this thickness from the depth of the valve. Scribe this difference onto your studs. The front of your blocks must line up with this mark.

4 Begin building the valve assembly. Because the assembly installed here has female threaded brass inlets and outlets, we soldered ½ in. threaded male transition fittings onto the one end of each tube entering and exiting the valve body. In the photo above, transition fittings have been sweated onto lengths of copper tubing to attach to the hot and cold supply outlets. Another fitting is being sweated onto a longer length of tubing to make the shower riser. Another tube and transition fitting are waiting their turn to become the spout supply tube. See the previous section on plumbing skills for information on making the connections you need to make for your assembly.

5 Attach the faucet/diverter valve body to a riser pipe. Apply Teflon tape clockwise to the male side of the fitting and apply pipe joint compound to the female section. Also attach the valve body to the supply tubes.

6 Position the valve/tubing assembly against the blocking in the wet wall. When the valves and riser pipe are at the correct height (it helps if you mark their height on the studs or blocking), use a marker to make cutting lines. Cut the supply tubes where they meet the existing supply tubes connected to the shutoff valves. Also mark the spout supply tube coming out of the valve body for connecting with an elbow fitting that transitions to the spout nipple. Make sure to allow a little space for the couplings that will connect the tubes.

7 Cut the tubing at your cutting lines. Attach a 90° elbow fitting to the top of the shower riser to transition to the showerhead nipple. Position the assembly in the wet wall and attach to the blocking with pipe straps. Then, join the supply tubes by soldering, using a repair coupling to make the joint.

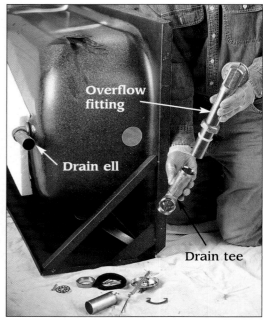

8 Temporarily put together the parts of the drain-waste-overflow (DWO) assembly so you can cut tubes and make adjustments as needed. Attach the drain ell to the tub with the strainer, then use slip nuts to connect the drain tee and overflow fitting so they roughly align with the drain ell and overflow opening when joined.

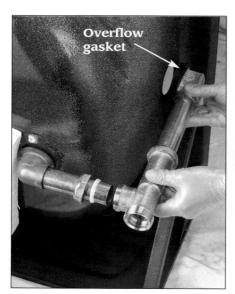

9 Attach the overflow gasket to the overflow fitting, then position the overflow ell precisely over the overflow opening. Align the drain tee with the end of the drain ell. Mark tubes for cutting as needed. Disassemble the parts and cut to fit according to your markings.

10 Apply pipe joint compound to both sides of the tub shoe washer and to the flange threads. Position the shoe and shoe washer under the flange for the drain strainer body. On the tub side, seat the flange into plumbers putty. Hold the shoe steady from below and twist the strainer to tighten it, using a screwdriver and open-end wrench (inset photo).

11 Coat the overflow gasket with pipe-joint compound. Affix the overflow flange and gasket to the overflow opening with the screws, nut or retaining plate (inset photo). Tighten the slip nuts with slip-jaw pliers and screw the retainer plate to the overflow pipe.

12 Install the overflow hood and linkage into the overflow ell and secure with screws. With a helper, position the tub (carefully) into the alcove. You'll be removing the tub once you have determined exactly where it needs to be in the alcove to align with the plumbing.

Access opening

13 Adjust the position of the tub until the drain tee on the DWO assembly is aligned with the drain trap opening in the floor. Make the connection, loosely, with a slip nut.

14 Make sure the tub is level and square in the opening. Mark the top of the tub flange on each wall stud with a marker to make reference marks for installing the ledgers that support the tub flange. If there is a gap between the flange and the wall studs, measure it so you can attach furring strips to the studs.

15 If you're building a stub wall or end wall for the alcove, transfer the tub position onto the floor and wall so you know where to build the new wall.

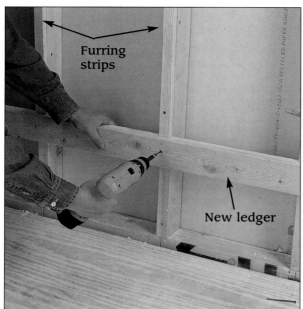

Furring strips

New ledger

16 Remove the tub and attach furring strips to studs as needed. Measure the height of the tub rim flange—from the top edge to the bottom of the rim. Measure down this distance from your marks on the studs and make new marks. Attach a ledger so the top is at the height of the bottom of the tub rim. Keep the ledger clear of the ends of the tub area by a few inches.

17 (OPTIONAL) Build a 2 × 4 partition wall if you're creating an alcove for the new tub. The outer stud in the wall should be flush with the tub apron. Nail or screw the sole plate to the subfloor so it will be aligned with the foot of the tub. Use a level bound to a straight 2 × 4 to transfer the location of the sole plate onto the ceiling so you can position a cap plate that's aligned with it. Attach the cap plate at ceiling joist locations, then nail the end studs and intermediate studs into position.

18 To reduce noise and conserve heat, wrap the tub with unfaced fiberglass insulation batting before installation. Secure the batting with string or twine.

19 Bathtubs (except claw-foot tubs) should be set into a thin mortar bed to prevent them from moving when they're filled with water, which can cause plumbing seals to fail. Trowel a layer of thinset mortar about ½ in. thick into the tub area using a square-notched trowel. The mortar also allows you to level the tub more accurately.

20 Lay a couple of 2 × 4s over the mortar bed to function as glides so you can slip the tub into position without disturbing the mortar. Rest the ends of the 2 × 4s on the sole plate of the wall. Slide the tub into place then remove the glides. Climb into the tub (with a surface protector, such as cardboard, in place) so the rim rests firmly on the ledger. NOTE: Make sure the slip nut and washer are slipped onto the drain tailpiece before installing the tub, and double-check to make sure the tailpiece is centered over the drain opening.

21 While standing in the tub, drive 1¼ in. screws with rubber washers into the wall studs above the flange. The washers should overlap the flange, pinning it in place. Don't screw through the flange. Allow the mortar bed to cure.

22 Attach strips of galvanized metal flashing around the perimeter of the tub, leaving a gap of about ¼ in. between the bottom of the flashing strips and the tub rim. This simply provides additional insurance against water penetration into the wall. Apply caulk between the tub apron and the floor.

23 If you're installing a tileboard tub surround kit, patch the walls and cover the partition wall with greenboard (water-resistant wallboard) then install the kit following the manufacturer's instructions. Apply pipe-joint compound to threads on the drop eared 90s for the shower arm and tub spout, since these won't be accessible once the tub surround is installed.

24 Test-fit as many ½ in. brass nipples at the spout as is necessary to find the right fit. The right nipple will bottom out in the drop-eared 90 and present plenty of thread for the spout to thread onto. It won't be so long that that the base of the spout can't be tightened flush with the wall with moderate torque. When you have the right nipple, wrap both ends clockwise with Teflon tape. Screw the nipple into the spout. Completely fill the back of the spout with plumbers putty, mounded beyond the base of the spout (See page 112). Screw the nipple and spout into the wall. You can use a hammer handle in the spout if you need extra torque to align the spout, but be careful not to break the drop eared 90 in the wall. The base of the spout should be flush with the wall. Wipe off extra putty.

25 Apply pipe joint compound to the female threads of the showerhead and showerhead shutoff valve (if you have one). Wrap Teflon tape clockwise onto both ends of the shower arm and onto the male threads of the showerhead valve. Screw the arm into the wall until tight and plumb. Push or thread the escutcheons on the shower arm and valve stems. Screw the valve onto the arm, and screw the showerhead onto the valve.

26 Apply valve-stem grease to the valve stems, and put on the handles. Apply Loc-tite to the handle screw and tighten the screws into the handles. Turn the water back on to inspect for leaks and to adjust the pop-up stop. Insert the pop-up stop, and install the trip lever and linkage if you have not already done so. Adjust the lift rod until the stopper works well. Examine the DWO and trap for leaks when the tub is full and when the tub is draining. Examine the faucet and the pipes and fittings of the water supply.

27 If you have a plunger-style drain, remove the over-flow hood and linkage and adjust and tighten the adjustment nut against the clevis. Reinstall the linkage, sealing the joint with plumbers putty.

Installing shower stall kits

Most of the parts of a shower drain kit are assembled and attached to the base prior to installation, then simply locked in place over the drain pipe once the base is installed.

Shower bases (pans) sold today are almost always sold as kits that generally include parts for attaching the base to your drain pipe, as well as walls for a shower surround. The specific installation steps vary by manufacturer, especially where making the drain hookup is concerned. Generally, the parts of the drain kit are partially assembled and attached to the base then set over the drain pipe and locked into place. The illustration to the left shows an exploded view of the drain kit parts on the base unit we installed. It is fairly typical of kits.

Prepare for the new stall by removing the old unit and making sure the drain and supply plumbing are in good repair and positioned correctly (some stalls come with supply holes precut into the walls, but many are left uncut so you can position the holes as needed to fit your supply pipes). Most prefabricated shower stalls require that the walls and floor in the installation area be smooth and clean. Scrape off old residue and apply a coat or two of paint primer to the stall area. If the wallcovering extends all the way down to the floor, trim it back in the stall area, so the base can be snugged up against the wall studs.

HOW TO INSTALL A SHOWER BASE & STALL KIT

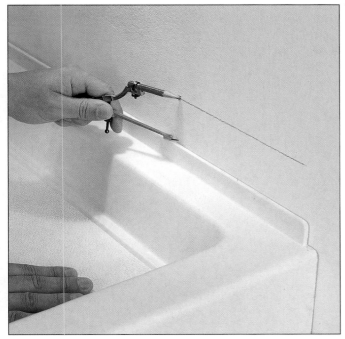

1 When installing a shower base in an area with finished walls (as opposed to new construction with no wall coverings installed), you'll need to cut away the wallcovering around the base so it can be attached directly to the studs. In most cases, make the cutting line about 3 in. above the top flange to create some wiggle room for fitting the base and drain kit assembly over the drain. Use a compass to scribe the profile of the curb onto the wall, then score along the line and pry off the wallcovering with a flat pry bar.

2 With the drain opening clear, set the base into the installation area and check it for level. In many case, you'll need to shim to bring the base to level (unless your unit is set into a bed of thinset mortar, as many are). TIP: Here's a surefire hint for testing the base for level: press a ¼ in. thick ring of plumber's putty around the drain opening to create a dam, then dump a couple of cups of water into the base. When the base is level, the water will form a concentric puddle around the opening, with no random pooling. To shim, use thin strips of exterior plywood or treated lumber (the strips should support the entire side of the base being shimmed).

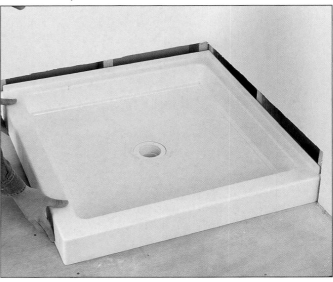

3 Assemble the parts of the drain kit according to the manufacturer's directions and attach it to the base (usually by tightening a lock nut on the under side of the base). Do not attach more parts than are recommended by the manufacturer. Most kits have a locking mechanism that is driven or twisted on the kit from above to fasten it to the drain pipe after the base is positioned and connected.

4 Lubricate the seal portion of the drain kit (the part that fits directly onto the drain pipe) according to the manufacturer's directions (warm, soapy water is a common lubricant). Make sure the lock nut on the underside of the base is fully tightened (but not overtightened), then position the base over the drain pipe. Set the base onto the pipe so the seal fits over the pipe.

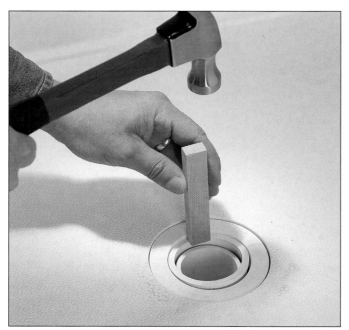

5 Lock the base into position. On the model we installed, this is done by driving a retainer wedge into the drain opening. This must be done with some care, as driving too hard can cause cracks or distension of the drain kit. After the fastener is secured (in our case, fully seated against the top of the drain pipe), attach the strainer to the drain assembly. If your shower base is set into mortar, let the mortar cure overnight before proceeding with the installation. NOTE: Some manufacturers recommend that you attach the flanges of the base to wall studs with either screws or nails. Unless it is specifically required, avoid doing this.

6 Mark layout lines on the wall for the surround panels, according to the manufacturer's directions. For the surround we installed, we marked vertical lines 30½ in. out from each corner and a level top line 68 in. up from the top of the base curb. Then, position the panels against the lines and tape them together temporarily with masking tape. Make sure that all overlaps between panels are within the range specified by the manufacturer (at least ½ in., but not more than 1 in., for the model shown here). If overlap exceeds the maximum, mark the panel that is overlapped and trim it to size by scoring with a sharp utility knife and snapping.

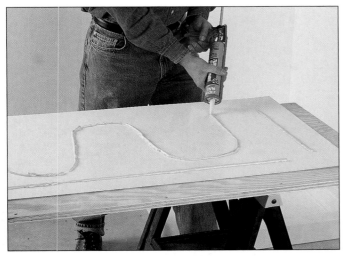

7 Attach the wall panel that doesn't contain cutouts for plumbing connections. Apply beads of panel adhesive formulated for bathroom panel installations to the back of the panel. The beads should be about ¼ in. in diameter and applied in a continuous "S" shape in the center of the panel. Apply straight lines of adhesive about 2 in. in from each edge. Press the panel into position, making sure to keep the bottom edge flush against the curb. Wrap a short piece of 2 × 4 with an old towel or cloth and use it to rub the panel to set it into the adhesive.

8 Mark cutouts on the wall panel that will fit over the plumbing stub-outs. If the cardboard shipping carton for the surround is in good repair, use it as a template for locating the holes. Cut the shower arm hole with a hole saw sized no more than ¼ in. larger than the diameter of the pipe. Use a utility knife to cut for the faucet assembly. Attach the panel, taking care to get a good seal with the first panel. If your surround has a corner panel, as ours does, install it after the plumbing wall panel. For alcove installations, install the third wall panel.

9 For corner units with no partition wall, install the support structure for the exposed shower wall (usually a clear, fixed panel). The kit we installed employs a system of U-shaped jambs that mount to the walls, and a frame network that supports the fixed panels from above and below. Finish installation of frame.

10 Complete the installation by hanging the shower door or doors. In the model shown here, a pair of narrow doors fit together in the front corner and slide apart next to the fixed panels to allow access to the shower. The doors are mounted with a system of pins and screws. Hang the doors then caulk the stall unit as directed. Attach the showerhead and faucet handles.

Toilets

The close-coupled toilet has a tank joined directly to the back of the bowl and is the most typical toilet design for residences. It uses gravity to flush wastes from the bowl into the drain. Low-profile toilets include the tank and bowl in one unit. Low-profile toilets are less efficient flushers than close coupled designs for the same reason that, on a given size river, a low waterfall is less powerful than a tall waterfall. We will be looking at close-coupled, gravity-flush toilets here.

Today's emphasis on water conservation means that close-coupled toilets have to use water more efficiently. Now most parts of the country require l.6 gallon toilet tanks, compared to tanks that held as much as 8 gallons a generation or two ago. The best low-flush toilets have tall, narrow tanks to take better advantage of gravity. Some low flush toilets let you adjust for a longer, slightly more voluminous flush by using an adjustable float on the flush valve chain than can be positioned to hold the flapper open longer.

Narrow, encrusted, or poorly vented drainpipes leading from the toilet will affect the performance of low-flush toilets even more than they affect the performance of older style toilets. For a low-volume flush to work, discharge from the toilet must meet with minimal resistance. An experienced drain clearing service can clear build-up from encrusted lines and remove blockages. If your drainage problems are chronic, consult a city building inspector or plumber to suggest modifications to your DWV system (See pages 51 to 53).

Low-volume, pressure-flushed tank toilets are also available now for residences. Like the "flushometer" toilets used in public buildings, pressure-flushed tank toilets blast out the bowl with pressurized water. While a flushometer toilet

Photo courtesy of American Standard

Two-piece toilets, also called "close-coupled" have a separate tank bolted to the back of the stool. This makes them easier to transport but also creates a potential problem area in the seal between the two parts—leaks can occur and moisture can get trapped in the seam. But all of the lower-priced toilets on the market are two-piece, making them considerably more common.

One-piece toilets, also called "low-profile" or "low-boy" generally are considered more attractive, and the absence of a mechanical joint between tank and stool makes them easier to clean and less prone to leakage. But they also tend to be weaker flushers and more expensive than two-piece fixtures. Parts also can be more expensive and difficult to find.

Exposed trap · Concealed trap

Concealed or exposed trap? Most toilets on the market today have a revealed trap: you can see the profile of the drain trap on the outside surface of the stool. For a sleeker appearance and ease of cleaning, some manufacturers produce smooth-sided toilets, but they tend to be more expensive.

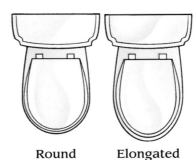

Round or elongated seat? The shape of the bowl determines the shape of the seat. If you have space and a few extra dollars to spend, consider a toilet with an elongated bowl and seat for greater comfort and enhanced sanitation.

Round · Elongated

Adding oomph to the flush

Assisted-flush toilets have a pressurized tank that compresses air when it fills with water. The compressed air forces the water out of the tank into the bowl with greater force than an ordinary gravity-assisted tank. The increased flushing power minimizes the number of "multiple flush" experiences associated with smaller 1.6 gallon tanks. Assisted-flush tanks are louder and costlier.

Gravity flush toilets are the standard. They rely solely on the power of gravity to draw tank water into the bowl, creating suction as it courses down into the trap. Generally, the taller the tank (relative to the bowl), the more flushing power the toilet has.

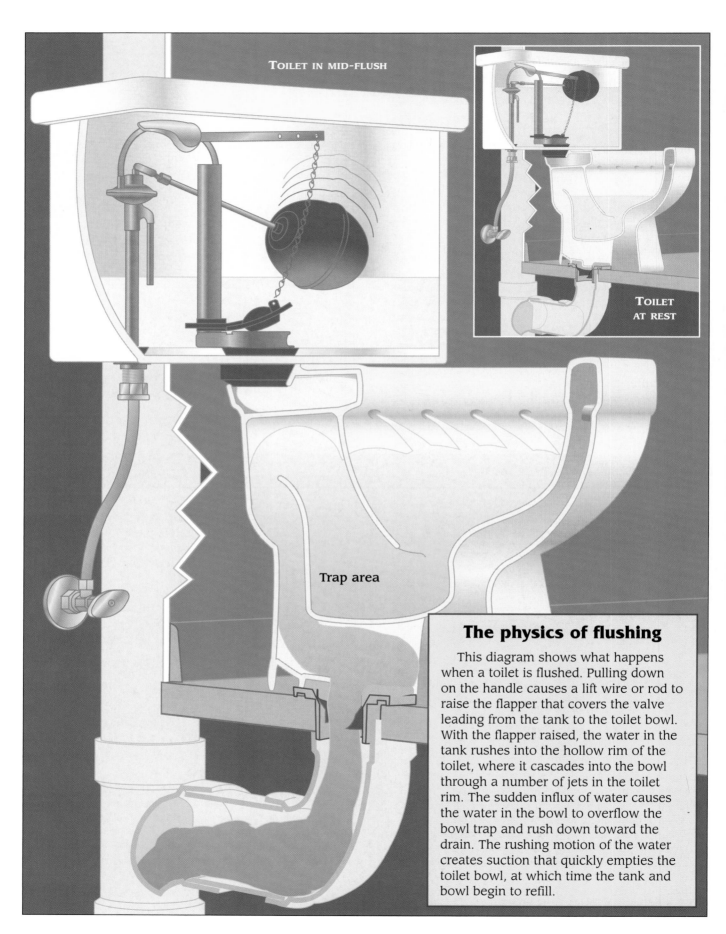

TOILET IN MID-FLUSH

TOILET AT REST

Trap area

The physics of flushing

This diagram shows what happens when a toilet is flushed. Pulling down on the handle causes a lift wire or rod to raise the flapper that covers the valve leading from the tank to the toilet bowl. With the flapper raised, the water in the tank rushes into the hollow rim of the toilet, where it cascades into the bowl through a number of jets in the toilet rim. The sudden influx of water causes the water in the bowl to overflow the bowl trap and rush down toward the drain. The rushing motion of the water creates suction that quickly empties the toilet bowl, at which time the tank and bowl begin to refill.

taps the water-supply pressure directly, a pressure flushed tank toilet stores pressure from the water system in a tank, making it look like a conventional close-coupled toilet. These toilets flush effectively with only 1.6 gallons of water. However, pressure-flushed tank toilets are expensive and are prone to failures that are expensive to fix.

How toilets work. The diagrams on these two pages show what happens when a toilet is flushed. Turning the handle lifts a flapper or tank ball at the bottom of the tank. Water rushes from the tank through holes in the rim of the bowl and a hole near the bottom of the bowl called the *siphon jet,* raising the water level in the bowl. The rising water in the trap area behind the bowl overflows toward the drain. This falling water creates suction that drags the contents of the bowl into the drain after it.

When the tank is empty, the flapper or tank ball flops back into place, cutting off flow from the tank to the bowl. The ballcock valve sends water through two tubes at this point: one directs water into the tank and the other directs water into the bowl through the overflow pipe. This second tube is called the refill tube; it ensures that the bowl is left with enough water.

When the water in the tank rises to a preset level, the ballcock valve turns off. Different ballcocks use different devices to shut down the ballcock valve. In the most familiar design, a float ball attached to a long float arm levers a plunger or diaphragm compression valve shut as it rises on the water. In another popular design, a float cup rides up the ballcock shank as the tank fills, levering the valve shut with a pull rod and valve arm. A floatless ballcock uses a pressure-sensing device to shut off the ballcock.

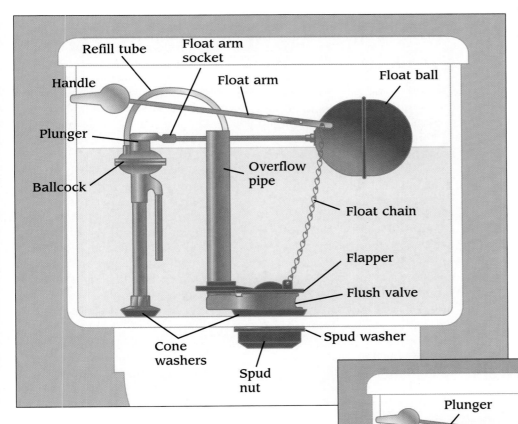

Float ball flush-&-fill mechanism

Most older toilets in use today employ this familiar system for flushing and filling the toilet tank. A float ball is connected to a plunger with a metal arm. When the toilet is flushed, the float ball sinks down to the lower water level, then gradually returns upward as the water level rises. When the tank is full, the float arm depresses a plunger on top of the ballcock, causing the inlet valve to close.

Float cup flush-&-fill mechanism

This variation of the float ball mechanism is becoming increasingly popular because it's more reliable than the float ball system. The basic principle is the same, but instead of a detached float, a plastic float cup mounted on the shank of the ballcock moves up and down with the tank water level. A pull rod connected directly to the plunger controls the flow of water into the tank. It also comes with an anti-siphon valve, which is now required by code for new construction toilet installations.

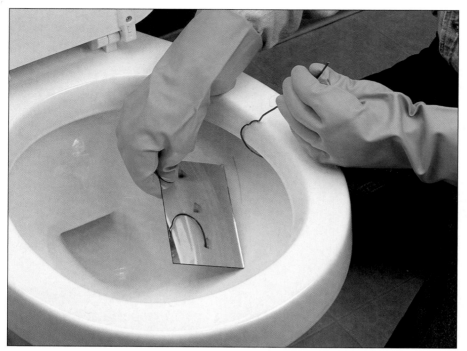

Weak flushes are aggravating problems that, in most cases, can be easily remedied. If adjusting the flush valve to allow more water to enter from the tank to the bowl doesn't strengthen the flush, the water jets in the bowl rim may be blocked. Clear lime deposits from the holes under the rim using a small mirror for direction and a coat hanger as a scraper. Also clear the flush valve hole in the bottom of the tank. Parts of old tank mechanisms may become lodged here. Use a commercial anti-lime compound to help break up and flush away hidden lime deposits. These usually require you to bail the bowl and pour an acidic solution down the overflow pipe of the tank. Enzyme build-up removers can help open up crusty toilet traps and the drain below the toilet.

Fixing Toilets

Toilet problems usually fall into four main categories: clogs, leaks to bowl (running), leaks to floor, and flush difficulties. You will need to take the lid off the tank to diagnose most of these problems.

Running toilets. Running toilets have problems with either the ballcock mechanism or the flush valve mechanism. The ballcock regulates fresh water flow to the tank, and helps fill the bowl after a flush. A faulty ballcock can continue to run, letting water flow through the overflow tube into the bowl and down the drain. The flush valve releases water to the bowl when you flush a toilet. If it leaks, water will also run continuously into the bowl. Ballcocks and flush valves can be entirely and inexpensively replaced. Keep this in mind before wasting too much time agonizing over their repair.

If your float works fine but your ballcock valve is still not shutting off the water, you will need to fix the valve or replace the whole ballcock assembly. Failed stem washers, seals or diaphragms can lead to continuous leaks. Failure of the washers or O-rings ringing the

stem can allow water to squirt up at the inside of the tank lid during tank refill. Turn off your water at the stop valve, flush the toilet, and disassemble the ballcock. Ballcocks can be replaced for less than five dollars and they don't require removing the tank to replace. This may be easier than locating parts for your ballcock.

The flush valve opens to let water flow into the bowl and closes to allow the tank to refill. If the flush valve doesn't close tightly, water will continue to run into the bowl and the ballcock will stay on or come on periodically. By opening up the back of the toilet, you usually can tell if the flush valve or the ballcock is your problem. Ballcock problems cause water to run into the overflow. With flush valve problems, the toilet will be running even though the water level is below the inlet to the overflow pipe. Flush valves use two kinds of stoppers at the bottom of the tank: flappers and tank balls. Both of these can develop leaks if they come out of alignment with the flush valve opening, if they wear out, or if the valve seat is damaged.

Weak flushes. Flushes that lack "oomph" may not be delivering

water from the tank to the bowl quickly enough or in great enough quantity. Blame the ballcock, the flush valve, the handle lever and chain, or the water inlet holes in the bowl. The ballcock is responsible for delivering enough water to the tank and bowl after a flush to make the next flush possible. The flush valve must let almost all the water drain from the tank before it closes again. The handle lever and chain must lift the flapper or tank ball when they are depressed but not interfere with the closing of the flapper or tank ball. The rim holes and siphon-jet hole in the bowl must be free of clogs that could keep water from flowing into the bowl quickly enough.

A flush valve that shuts prematurely or a tank that doesn't fill to the fill line will deliver weak flushes. Partially blocked flush holes on the rim of the toilet or a partially blocked siphon-jet hole in the bowl can slow and weaken the flush. A partially clogged or encrusted toilet trap or drain can produce a weak flush as well.

If a flush valve doesn't stay open or closes prematurely, try taking slack out of a loose chain or adding slack to a tight chain. The trip lever should only lift ½ in. before it unstops the flapper. You may adjust some flappers to stay open longer by moving a mini float down the lift chain. Replace a flapper that won't operate properly.

Too little flush water can produce a weak flush. Adjust the float or pressure sensor on the ballcock if your tank doesn't fill to within ½ in. of the opening to the overflow pipe. Floats should be moved upward by bending the float arm slightly or, with shaft-mounted floats, sliding the float up the pull rod a little by depressing the spring clip. With floatless ballcocks, turn the adjustment screw clockwise a quarter turn. If the bowl seems low on water between flushes, make sure the bowl refill tube is directing water into the overflow pipe after flushing. Reconnect disconnected refill tubes. Clean or replace a ballcock valve that's not sending water through this tube

Leaks. Leaks out of the toilet can happen when the toilet bowl or tank is cracked or defective, or when one of the seals is broken. The wax ring sealing the juncture of the toilet bowl and the toilet flange can fail, causing water to seep from under the bowl. The washers sealing the flush valve to the tank and sealing the junction between the tank and the bowl can leak. The holes for the two or three bolts holding the tank to the bowl and the hole for the tailpiece of the ballcock can leak. The supply tube fittings can leak. Sometimes a faulty ballcock can squirt water at the lid after a flush, which then trickles down outside the tank. Finally, cold water in a tank can cause water from the air to condense on the tank surface. This can be misinterpreted as a leak.

Closet risers. Closet risers are supply tubes that connect the shut-off valve to the ballcock. These are similar to sink and lavatory supply tubes, except the toilet end uses a different nosepiece configuration. As with sinks, braided stainless steel or chromed brass connectors are a good choice. Usually, your toilet will use a ⅜ in. outside diameter tube. If you need to replace a closet riser or add/replace a supply stop, see pages 44 to 49. Use closet risers for supply tubes.

STEPS FOR CLEARING A CLOGGED TOILET

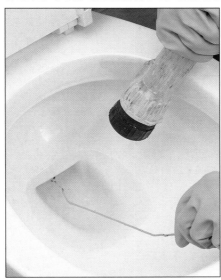

Visible clog in trap. An object that falls in the bowl and lodges in the trap can create a blockage. Use a small mirror and a flashlight to see if you can see the obstruction. A wire coat hanger with a hook bent on one end or mechanical fingers can be used to retrieve some objects. In some cases, you can use a closet auger to try and retrieve cloth items like diapers.

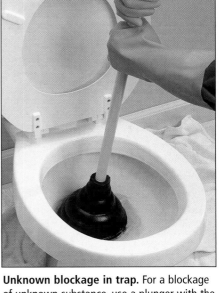

Unknown blockage in trap. For a blockage of unknown substance, use a plunger with the skirt folded out and pump hard and fast against the trap opening. Try ten to 15 cycles of five strokes each. You need water in the bowl to make the suction. Manually lift the flapper or tank ball to let water into the bowl, but put it down again before the bowl fills too much.

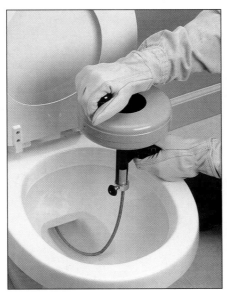

Stubborn blockages. Use a hand auger if the plunger doesn't work. Turn the handle clockwise while pushing the cable back and up into the trap opening. You will need to work to get the auger through the tight turns in the trap. When the cable is fully extended, crank and pull the cable back in. If the flush is still sluggish, run the auger through the trap again, emphasizing first one side and then the other side of the trap. The closet auger can also retrieve cloth items stuck in the trap.

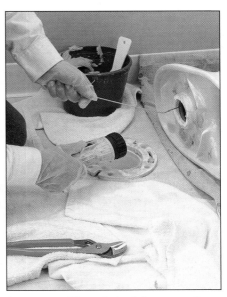

Last resort. When all else fails, remove the bowl and attack the clog from the opposite side of the trap area. Roll the bowl into different positions to get small solid objects to fall out. You can take the toilet outside and flush it out with a garden hose to clear cat litter or other hard-to-remove material. Wrap electric tape around the metal hose threads to keep from damaging the porcelain (to remove your bowl, see "Replacing a toilet & wax ring").

Toilets 139

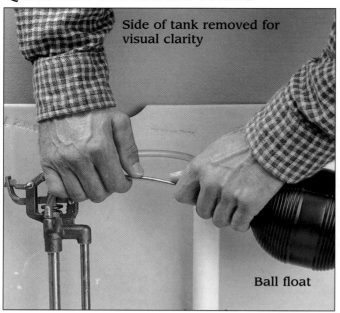

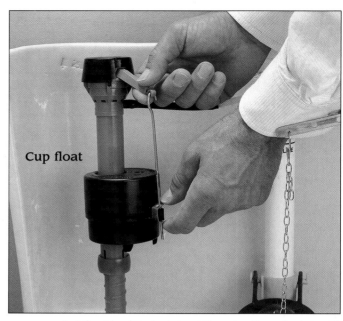

Adjust the float. Lift off the lid on the toilet tank. If you see water flowing into the overflow inlet, your float ball or float cup is not closing the ballcock valve. If the mechanism has a ball-style float (left photo), bend the float-ball rod downward a little, so the rod closes the ballcock before the water in the tank reaches the overflow level. If the ball scrapes against the side of the tank, bend the rod so it can float clear. If the mechanism has a cup float (right photo), compress the clip and slide the cup a little farther down the rod. If you are trying to increase water level in the tank, adjust floats in the opposite direction.

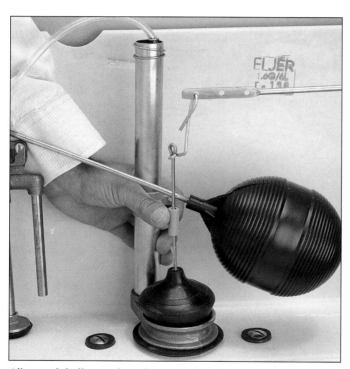

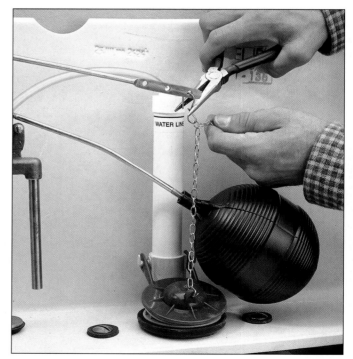

Align tank ball over the valve seat. The lift wire of a tank ball is held in alignment with an adjustable bracket. Position the bracket and wire so the ball drops plumb into the valve seat. You may have to straighten or replace a bent wire.

Adjust chain on flapper. Flappers that don't stay open when the flush handle is turned may have too much slack in the chain connecting the flapper to the trip lever. Take out slack until the trip lever can only move about ½ in. before pulling on the flapper. Flappers that won't drop into the closed position after a flush may have extra chain interfering with the flapper. Remove extra links that may be unbalancing the flapper. If this doesn't work, replace the flapper.

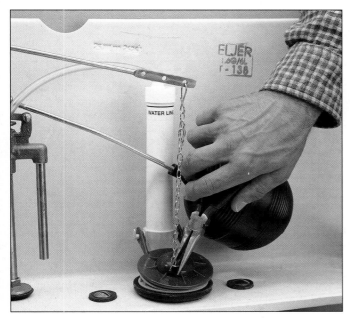

Reconnect the flapper or float. The toilet handle lifts a lever that pulls up on a chain or wire connected to the tank flapper or ball. A loose chain may not lift the flapper high enough, requiring the person flushing to hold the lever down through the flush. Or the toilet may flush fine, but then continue to run. This may be the fault of the flapper or tank ball, or the handle mechanism could be keeping the flush valve from reseating properly. A flush handle that does nothing when depressed might be disconnected.

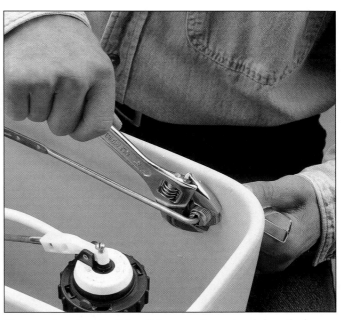

Fix or replace handle. Take off the mounting nut and clean the handle lever and socket with a wire brush and vinegar. The mounting nut of the handle loosens counterclockwise, just the opposite of a typical nut. Replace a damaged flush lever.

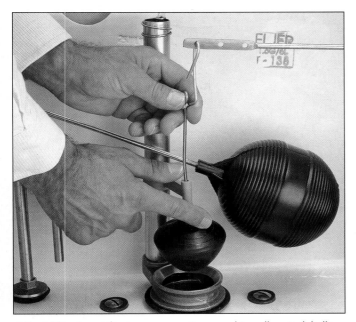

Adjust lift wires. Lift wires are sometimes used to pull up tank balls, which then must float back down into the flush valve seat when the tank empties without interference from the lift wire. Adjust lift wires so they operate without sticking. The wire connected to the tank ball must be straight so it can slide through the guide arm easily.

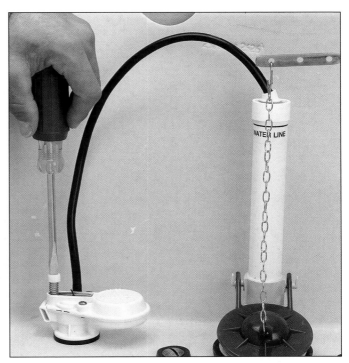

Adjusting a floatless ballcock. The floatless ballcock senses water pressure. Lower and raise the water level by turning the adjustment screw a quarter to a half turn, counterclockwise or clockwise respectively. Floatless ballcocks are no longer allowed for new installations under most codes.

CORRECTING TANK LEAKS

Leaks between the tank and bowl. Sometimes flushing the toilet will reveal water leaking from the flush valve area between the tank and bowl (or it may leak whether you're flushing the toilet or not). The cone or spud washer may be leaking, or the bolts holding the tank to the bowl may be leaking. If you can rock the tank on the bowl, try tightening the tank. Take the lid off and place a level across the tank. Try leaning the tank toward level. Holding the appropriate screw in the tank with a large screwdriver, snug up the up-slope bolt and then the other bolt with your fingers or a socket wrench. (Never turn the screw on the rubber washers.) Tighten the nuts a little until the tank is level and stable but not beyond this point, as you could squeeze the rubber washers from under the bolts or crack the tank. If tightening nuts doesn't stem the leak, replace washers at the tank bolts and at the flush valve tailpiece.

Tracking tank leaks

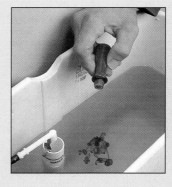

Water collecting behind the toilet seat or on the floor may be coming from the tank. Dry the water and flush the toilet repeatedly to determine where the water is coming from. Sometimes a faulty ballcock will jet water at the inside of the lid, and it will run down the outside of the tank. Cracks in the porcelain require that you replace the tank. Otherwise, fix a leak from one of the four openings in the bottom of the tank. If you don't know where the water is coming from, put food coloring in the tank when you won't be flushing for some time. If color ends up on the floor, and not in the bowl, one of the washers or seals below the tank is leaking, not the bowl. If you suspect water is leaking from the bowl, put food coloring in the bowl. See if the water collecting around the toilet is colored. A cracked bowl might spill the color between flushes. When a wax gasket fails, the color would escape during the flush. If you've ruled out the possibility that water collecting on the floor is coming from the tank, then you must research how it is coming from the bowl. Water leaking at a constant rate from the bowl, regardless of how much the toilet is used, indicates that the bowl or its trap (part of the bowl) is cracked and needs to be replaced. Water that seeps from under the toilet during or shortly after a flush indicates that the wax ring sealing the toilet to the toilet flange is leaking. Replace the wax ring.

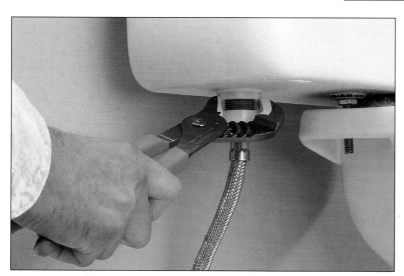

Fixing a leak from the base of the ballcock. Water leaking from the ballcock tailpiece area may result from an imperfectly sealed ballcock where the cone or grooved washer fits into the hole on the inside of the tank. Or the union between the supply tube and the ballcock tailpiece may be leaking. Try tightening the supply-tube coupling nut or the ballcock mounting nut a quarter turn. If the leak continues, you may tighten the appropriate nut a little more, but you probably need to replace the washer between the inside of the tank and the ballcock. Or, you may need to replace the supply tube. Exercise restraint when tightening any nut or bolt on porcelain, as it can crack.

Control condensation on the toilet tank. If the tank seems to be sweating water and your tank-water temperature is significantly colder than the air, you probably have condensation rather than a leak from the tank. Insulating your tank can control this. Buy a toilet liner kit or use ½ in. rigid foam to make your own liner. Turn off the water, flush the toilet, and scrub out the inside of the tank with an abrasive cleanser. Rinse and dry the tank then use water-resistant adhesive to attach the insulation to the inside walls of the tank. Let the adhesive cure as directed.

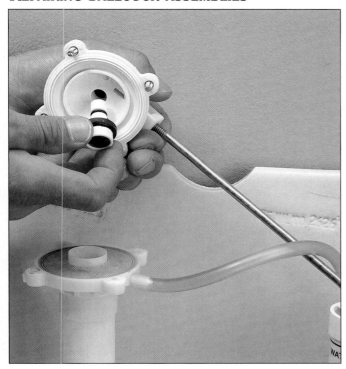

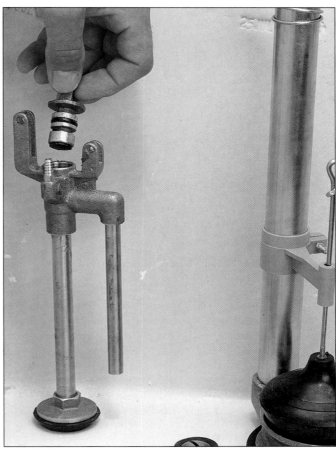

Fixing a diaphragm ballcock. Remove the four screws from the top of a diaphragm ballcock and lift off the bonnet. Use an abrasive scrubber to remove buildup on the inside of the valve. Remove and replace worn neoprene parts on the plunger and replace the diaphragm. Coat these neoprene parts with heatproof grease before installing. Replace the entire ballcock if necessary.

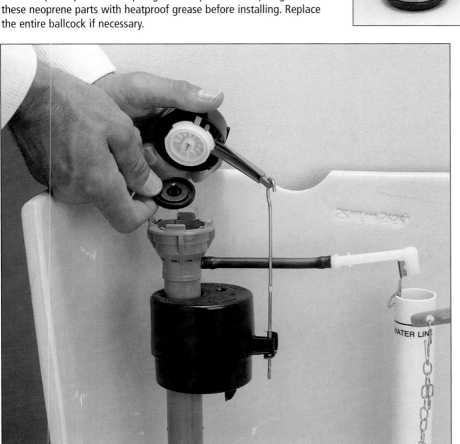

Fixing a plunger ballcock (above). Access this old-fashioned ballcock by removing thumbscrews attaching the float rod to the ballcock. Set the ball and rod aside. Pull out the plunger. Replace the seat washer at the bottom of the plunger and any other washers or O-rings you find. Coat these parts with heatproof grease before installing. Sometimes old leather washers will ring the stem to keep water from spraying up through the top of the valve. Replace these with neoprene substitutes. Clean out sediment inside the ballcock with a wire brush. If this doesn't do the trick, replace the mechanism with a more modern diaphragm or float-cup ballcock.

Fixing a float-cup ballcock (left). Pry off the cap and remove the bonnet by lifting the float lever and pushing the mechanism down while twisting counterclockwise (like opening a bottle of prescription medication). Clean out the mechanism with a wire brush and replace the seal that fits under the lever mechanism if this looks worn. Alternatively, replace the entire ballcock.

HOW TO REPLACE A BALLCOCK

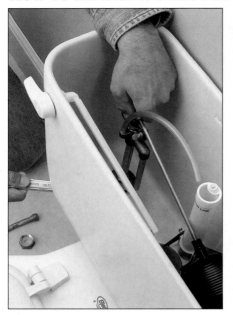

1 Shut off the water. Flush the toilet, and sponge any remaining water out of the tank. Disconnect the coupling nut holding the supply tube to the ballcock tailpiece. Next, clamp 6-in. locking pliers on the base of the ballcock in the tank. Remove the mounting nut below the tank with slide-jaw pliers. Remove the old ballcock.

2 Coat a new cone washer with pipe-joint compound for added insurance, and push it onto the new ballcock tailpiece so the beveled side will press against the hole in the tank. Insert the tailpiece into the hole. Attach the float arm and float for float-ball models.

3 Insert the refill tube into the socket provided in the overflow pipe. It may clip into the overflow pipe on some models. Trim the tube with scissors if necessary.

4 Screw the mounting nut and the supply-tube coupling nut onto the ballcock tailpiece. Before tightening the mounting nut all the way, align the ballcock in the tank so the float ball will not rub against the tank (for float-ball models). Gently tighten the mounting nut and the supply-tube coupling nut with your adjustable wrench. Turn on the water and check for leaks. Tighten the mounting nut or coupling nut a quarter turn more if water leaks past either seal.

5 Flush out mineral debris. Turn and lift the cap of the ballcock and turn the water on at the shut-off. Use the cap to keep water from escaping the tank. Turn the water off and on again to make sure all debris escapes before securing the ballcock cap in place again. Adjust the ballcock to turn off when the water level in the tank is about ½ in. below the overflow pipe opening. On shank-mounted float cup models, depress a clip to move the float cup up or down the pull rod. On float ball models, bend the float arm up or down to adjust the water level.

TIPS FOR ADJUSTING FLUSH VALVES

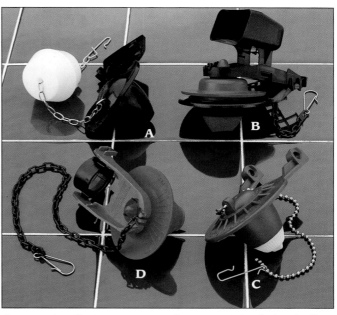

Replace flappers and tank balls that can't be made to sit tight. Flappers unhook from lugs on the overflow pipe and tank balls unscrew from their lift wires. If the flush ball guide or lift mechanism is damaged, replace the ball mechanism with a more reliable tilt-back flush mechanism.

Conservation Tip: To save water, replace your existing flush valve flapper with a variable buoyancy flapper or another of the water saving flappers on the market that let you adjust how much water flows from the tank during a flush. Shown above are: (A) Adjustable float flapper for ease of water-level adjustment; (B) Timing cup with drain holes to control water usage; (C) Dial flapper for setting water usage level; (D) End-cap dial flapper.

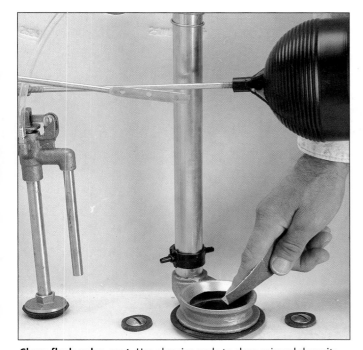

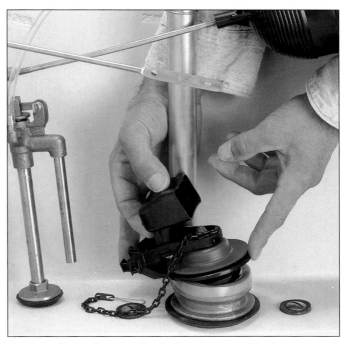

Clean flush valve seat. Use abrasive pads to clean mineral deposits from a plastic valve seat and emery cloth to clean a brass valve seat. Sometimes the valve seat will become pitted or corroded and need replacement. A flexible valve seat may be pried out and replaced. Otherwise, you may replace the flush valve in its entirety. This can mean a lot of work with a hacksaw if the bolts are corroded in place.

TIP: Buy a flush valve repair kit. Replace the seat and its flapper or tank ball with a replacement repair kit. This sticks in place over the existing valve seat, which saves you the trouble of having to remove the tank.

HOW TO REPLACE A FLUSH VALVE

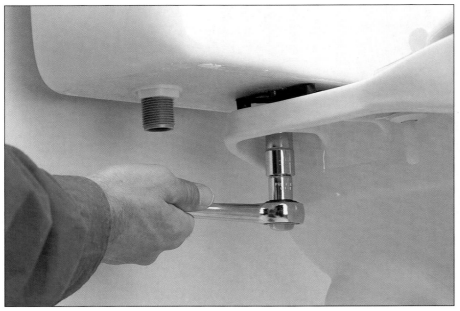

1 Shut off the water. Flush the toilet and sponge any remaining water out of the tank. Disconnect the coupling nut holding the supply tube to the ballcock tailpiece. Unscrew the nuts from the tank bolts with a deep socket wrench. Hold the screws inside the tank with a large screwdriver. Use penetrating oil to loosen corroded nuts, or remove the toilet seat and cut the bolts between the tank and the bowl with a hacksaw. Take the tank off and set it aside. Note: a second set of nuts between tank and bowl may hold the tank bolts to the tank. Cutting the tank off may mean cutting through these nuts as well as the bolts. Try sincerely and with the right tools to unthread the lower nuts from the tank bolts before subjecting yourself to this hacksaw experience.

2 Use a spud wrench or large slide-jaw pliers to remove the spud nut holding the old flush valve to the tank.

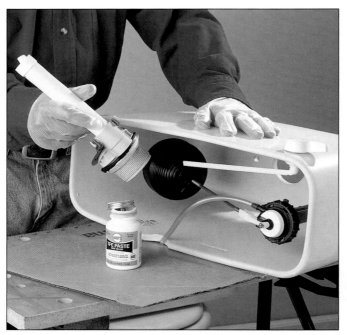

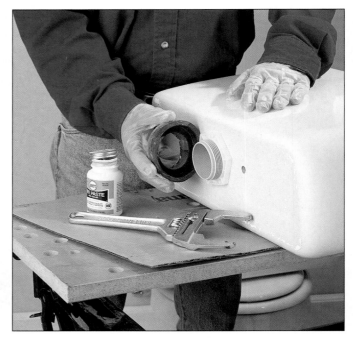

3 Fit a large cone washer flat-side first over the tailpiece of the flush valve. The beveled side will press against the hole in the tank. Position the valve in the tank so the overflow pipe faces back. If necessary, cut the overflow pipe so the opening falls at least an inch below the hole in the tank for the flush handle. It must also open below the critical level marked on the ballcock.

4 Thread the spud nut onto the tailpiece and snug it up with your wrench or pliers. A fiber or steel washer may go on next. Lastly, place the large soft spud washer over the tailpiece. Coat the spud washer with pipe joint compound where it will contact the water inlet hole in the bowl.

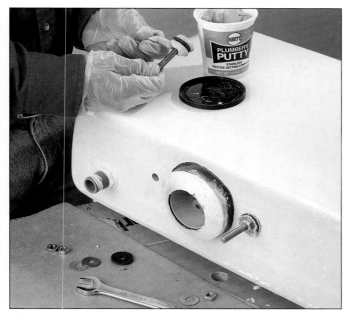

5 Apply rings of plumbers putty under the heads of the tank bolts then slip the rubber washers onto the bolts. Add rings of plumbers putty under the washers and against the shafts of the bolts. Insert the bolts through the inside of the tank. Place a rubber washer, metal washer, and nut on each bolt protruding through the bottom of the tank.

6 Flip the tank right side up and insert the tank bolts into the holes on the back of the bowl. Put metal washers and nuts on the tank bolts projecting under the shelf of the bowl, and hand-tighten these. Place a level across the top of the tank. Carefully snug up the bolts with your fingers a little at a time, alternating between nuts, pressing the tank toward level in the process. The nuts should not be overtightened. On many toilets, the tank pivots on the spud washer and a gap between the tank and the bowl will remain.

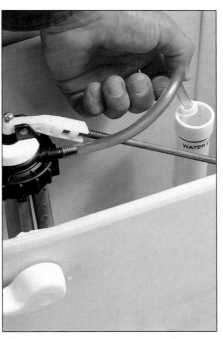

7 Attach the water supply tube. Tighten the supply nut a quarter-turn past hand tight with an adjustable wrench. It's better to have it too loose and have to give it an extra quarter turn later than to strip the plastic threads of the tailpiece.

8 Attach the chain between the flapper and the toilet handle arm. The handle lever should move only about ½ in. before the slack is taken out of the chain. Reposition clip on the chain with needle nose pliers and remove extra chain.

9 Attach the refill tube to the overflow pipe. You may remove the overflow pipe ring if the refill tube clip requires a bare pipe. Open the shutoff valve. Check for leaks and check the flush.

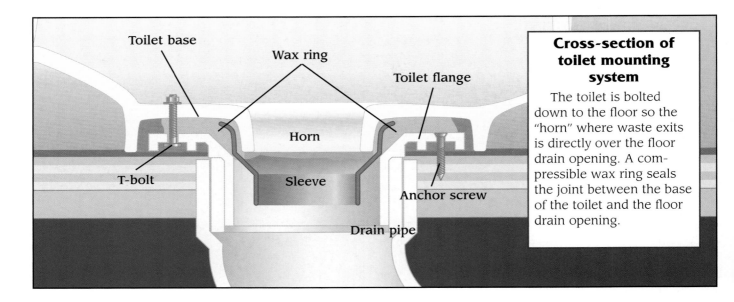

Toilet base

Wax ring

Toilet flange

Horn

T-bolt

Sleeve

Anchor screw

Drain pipe

Cross-section of toilet mounting system

The toilet is bolted down to the floor so the "horn" where waste exits is directly over the floor drain opening. A compressible wax ring seals the joint between the base of the toilet and the floor drain opening.

Replacing a toilet & wax ring

Replacing a toilet is not as difficult as many people think—as long as the closet flange and floor around the base of the old toilet are in good shape. Once you have removed the old fixture, inspect these areas closely for signs of damage or rot. If you spot any problems, you'll definitely want to take care of them before installing the new toilet.

When buying a new toilet, don't simply rush out and get the cheapest one off the floor at your local building center. Certainly there are times when this approach might make sense, but consider your options. There are many factors to evaluate (See page 133).

Whichever style toilet you choose, make sure it will fit your space. The distance from the closet flange (drain) to the wall determines what size you should purchase. Measure the distance from the bolts to the wall. If there are four bolts, measure from the bolts that are attached to the flange. This "rough-in" distance determines which size toilet you must buy (12 in. is the minimum distance and most common these days).

Along with the toilet itself, you'll need a new wax ring, or its equivalent (See tip box, right). Most wax rings come with the plastic sleeve in the center preinstalled. You'll also need to purchase flush mechanisms in many cases, along with a toilet seat. The toilet manufacturer may provide closet bolts and other hardware with the unit. Inspect these. If they are made of plated steel, replace them with solid brass equivalents.

Options for raising a low toilet flange

Especially when remodeling and replacing a thin floorcovering (like sheet vinyl) with a thicker covering (like ceramic tile), you can end up with a floor surface that's higher than the drain flange. Below are two solutions to this common problem.

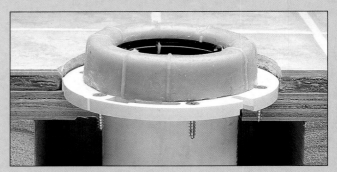

Use an extra-thick wax ring if the top of the current flange is level with or slightly below the surface of the floor that will contact the toilet base. This is a better method than stacking two regular wax rings together (a common practice, but not recommended).

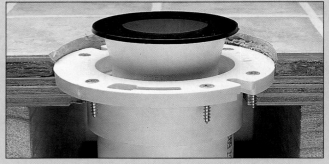

A waxless flange spacer ring can be installed on top of the original flange to form a seal around the toilet horn. Consisting of a plastic ring and flexible rubber gasket, a spacer ring assembly is a neater solution than a wax ring.

HOW TO REPLACE A TOILET

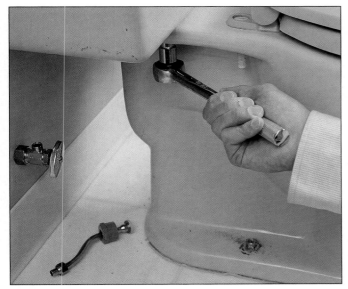

1 Shut off water to the toilet. Flush the toilet, then sponge out any water remaining in the tank and bowl. Remove the water supply tube with an adjustable wrench. Remove the nuts from the tank bolts with a socket wrench. You will need to hold the bolts steady with a screwdriver as you do this. Use penetrating oil on corroded or stuck nuts, or as a last resort, cut the bolts with a hacksaw between the bowl and tank. Lift the tank off the bowl and set it on towels. Pry off the plastic caps covering the floor bolts.

2 Remove the floor nuts with a socket wrench. If the bolts are stuck, try penetrating oil or cut the bolts with a hacksaw. Rock the toilet to break its seal with the toilet flange then lift it off the bolts and set it on its side. Some water may spill from the bowl or base. Scrape the old wax from the toilet flange on the floor with a putty knife, and then stuff a damp rag into the drain opening to keep sewer gasses out of the house. Replace the closet flange or use a flange repair kit if it is damaged. Scrape off any plumbers putty on the floor. Clean, and dry the floor. Scrape wax from the toilet if you intend to reuse it. Otherwise, dispose of it right away.

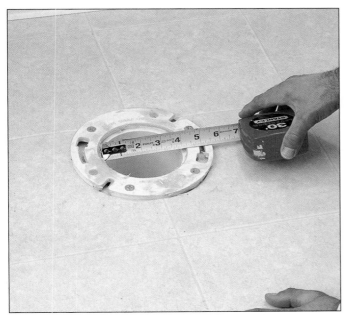

3 Measure the opening in the closet bend (drain) at this time. If it is 3 in., you should get a 4 × 3-in. wax ring. If it is 4 in., you will get a 4 × 4 ring, if you can find this size. Buy a ring with a sleeved "no-seep" plastic flange You may want to purchase new brass closet bolts, nuts and washers, too. The ones that come with the wax ring are often plated steel and can rust. Buy an extra set of washers and nuts so you can pre-attach the bolts to the closet flange.

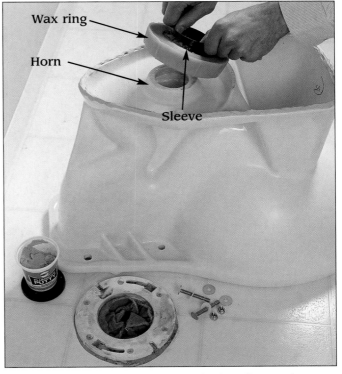

Wax ring

Horn

Sleeve

4 Center the closet bolts in the slots in the closet flange and secure them with a washer and nut. Put the new wax ring over the closet flange, with the funnel shaped plastic sleeve directed into the drain.

Fixing faulty flanges

Have a wobbly toilet? Your toilet should be secured to an intact closet flange that itself is securely attached the subfloor. If the flange screws have pulled loose due to minor rot or if the hole in the floor is too big for the closet flange, buy a repair flange kit (left photo). This consists of heavy-gauge slotted metal plates that slide under the flange and screw to the floor. The closet flange is then screwed to the repair flange

If your flange is corroded or broken, you may remove and replace it or screw a stamped metal ring

called a "Quick Ring" directly on top of the old flange (right photo). If you're tiling or increasing the thickness of your floor, you'll need to bridge the gap between your toilet and the closet flange (See page 148).

HOW TO REPLACE A TOILET (CONTINUED)

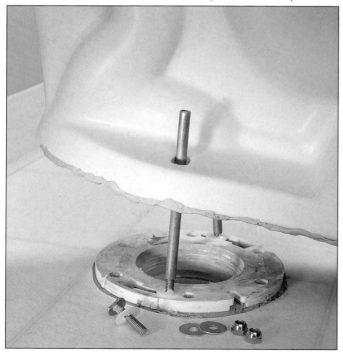

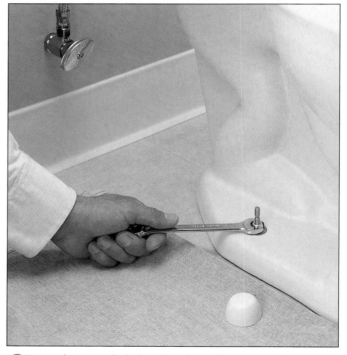

8 Press a bead of plumbers putty around the rim of the base of the toilet. (NOTE: some plumbers advise against this bead, since you won't be able to tell if your wax ring fails because the putty will keep leaked water trapped under the toilet). Lower the toilet into position over the floor bolts. We used lengths of copper tubing to extend the bolts, making it easier to align them with the holes in the stool. Remove the guides, if you use them, then press down on the bowl to seat it into the bed of plumbers putty.

9 Put washers over the bolt ends and snug the nuts onto the bolts. Tighten the bolts alternately until the toilet lies flat on the floor. Do not over-tighten. Cut off excess bolt ½ in. above nuts with a hacksaw. Fill plastic bolt covers with plumbers putty and push onto bolt ends.

10 Place a new spud washer over the tailpiece of the flush valve. Coat the spud washer with pipe joint compound where it will contact the water inlet hole in the bowl. Finish installing the flush valve (See pages 144 to 145).

11 Put rings of plumbers putty under the heads of the tank bolts, then put the rubber washers on the bolts. Put rings of plumbers putty under the washers and against the shafts of the bolts. Insert the bolts through the inside of the tank. Flip the tank right side up and insert the tank bolts into the holes on the back of the bowl as you lower the tank into position.

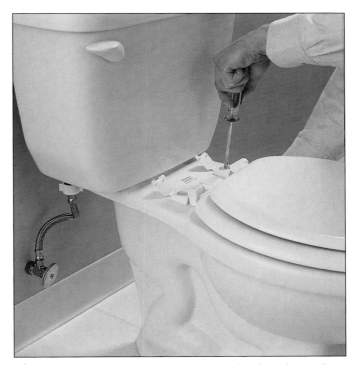

12 Put metal washers and nuts on the tank bolts projecting under the shelf of the bowl, and hand tighten these. Place a level across the top of the tank. Lean on the tank to level it, if necessary. Carefully snug up the bolts with a socket wrench or your fingers a little at a time, alternating between nuts, pressing the tank level in the process. The nuts should not be overtightened. Hold the tank bolts steady with a screwdriver from above while you tighten the nuts.

13 Attach the closet riser (water supply tube). Tighten the coupling nut a quarter turn past hand tight. It's better to have it too loose and give it an extra quarter turn later than to strip the plastic threads of the tailpiece. Install the toilet seat. Turn on water at the shutoff valve and test.

Water heaters

How water heaters work.

Cold water enters the tank under pressure through the dip tube. Hot water is released from the tank under pressure through the hot water outlet. A thermostat/temperature control senses when the water temperature drops below the preset temperature (usually between 120 and 140° F). In electric heaters, each of two heating elements has its own thermostat, which switches power on and off to the heating element it controls. In gas heaters, the single thermostat opens and shuts gas supply to the burner, where the gas is ignited automatically by the pilot light, which always stays on. A thermo-

couple senses if the pilot goes out and shuts down gas to the pilot and the burner. The flue in gas heaters conducts gases vertically through the tank and out the top vent. Both gas and electric heaters have anode rods, made of magnesium, that corrode away gradually through oxidation, preventing corrosion of the tank. Both have manually operated drain valves to draw off water and sediment from the bottom of the tank and temperature and pressure relief valves to allow water and steam to escape if pressure builds to a dangerous level.

General water heater maintenance.

• Drain a couple of gallons of hot water off the bottom of the tank through the drain valve every two months. Get in the habit of filling your mop bucket here occasionally. This will remove sediment from the bottom of the tank, which interferes with the ability of a gas burner to heat water and can shorten the life of any tank. Use the drain valve, not the pressure release valve, and be careful: the water here is extremely hot!

• Contact your water heater dealer to discover if there is a chemical de-liming procedure useful for your local water conditions. Lime and scale deposits can reduce the efficiency of your heater by 30% and shorten the life of the heater. Lime and scale buildup are not removed when the tank is drained.

• Operate the pressure release valve yearly. Lift the valve lever a number of times until the valve operates and reseats perfectly.

• Keep the thermostat set only as high as necessary to efficiently run your hot water appliances. This saves energy and extends the life of your tank.

Electric heater maintenance.

Most electric water heaters have a top and bottom element. If nothing else is wrong, "warm-water-only" indicates a burned out top-heating element. If the bottom-heating element burns out, you'll get hot water, but not very much of it. Replace the broken element. If this doesn't work, replace the problem element's thermostat. A plate on the side of the water heater will tell you the voltage and wattage of the parts. You can test the heater elements to see if they're working with a multimeter (an electrical testing device). See photo, left.

Gas heater maintenance.

• Remove the outer and inner panels below the thermostat box and take a look. Is the pilot light burning? If the flame has gone out, re-light it.

• If the pilot goes out continually or will not light at all, replace the thermocouple and clean the pilot gas tube.

• Clean the gas burner annually to improve the energy efficiency of the heater and help it last longer.

• If the burner flame is yellow and smoky, clean the burner jets and the burner gas tube nipple. Make sure the heater is getting sufficient combustion air. Provide a ventilation tube if needed.

• If you smell fumes from a gas water heater, or a carbon monoxide tester indicates high levels of the gas, make sure the vent hood and vent pipe of your water heater are clean and connected properly. The male ends of the pipe should always face away from the heater.

Leaks.

• *Hidden Leaks.* Low water temperature or low hot-water supply may result from a leak in your hot water supply piping, in a fixture or appliance, or at the heater through the pressure release valve if a drain line hides the spout. Drain lines on leaking valves will feel warm some distance from the valve. The existence of hidden leaks can be detected at the water meter. Mark the position of the dial and note

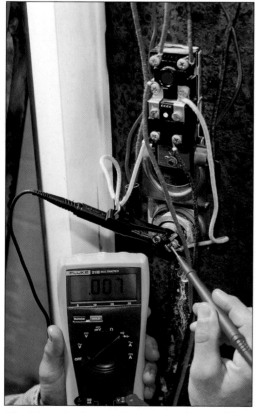

Electric water heater elements can be tested for "continuity" to determine if they are the cause of a nonfunctional water heater. Touch the probes of a multimeter to the element terminals. Make sure the multimeter is set to determine continuity and be sure the power to the heater is off, otherwise you could ruin the meter. You'll hear a buzz if the water heater element is in good working order.

the digits. See if they change over when no appliance or fixture is drawing water.

• *Visible Leaks.* Leaky water inlet/outlet nipples, relief valves or drain valves may be replaced or re-wrapped with Teflon tape. You will need to turn off the heater, shut off the supply valves, and drain down the tank before replacing a valve. A dripping relief valve may indicate that the water temperature is set too high or may indicate the need for a pressure-reducing valve at the water meter. Water at the vent hood or at the bottom of the heater may be condensation.

High water pressure or continual activation of the relief valve.

High water pressure in the tank can lead to activation of the pres-sure relief valve and premature wear of the heater and valves else-where in the house. Installing a pressure-reducing valve (See page 19) can extend the life of your heater and other appliances.

If you have a pressure reducing valve or a check valve (a fitting with a directional arrow) on the supply line to the heater, expand-ing hot water in your tank can't back up into the cold supply. To take pressure off your relief valve, install an expansion tank on the cold-water side of your heater, between the heater and the check valve. If your water inlet nipple has a directional arrow, it may have an integral check valve and should be replaced with a regular brass nip-ple. Install a new check valve on the other side of the expansion tank from the water heater.

Replacing a water heater.

Replace a water heater that leaks from within (water comes from underneath). Drain valves and pressure relief valves can be replaced. Gas burners, thermostats, and electric heating elements can be replaced. But once the tank has corroded through, your only option is to replace the whole water heater. Replacing a water heater is a relatively complicated project with potential for major problems.

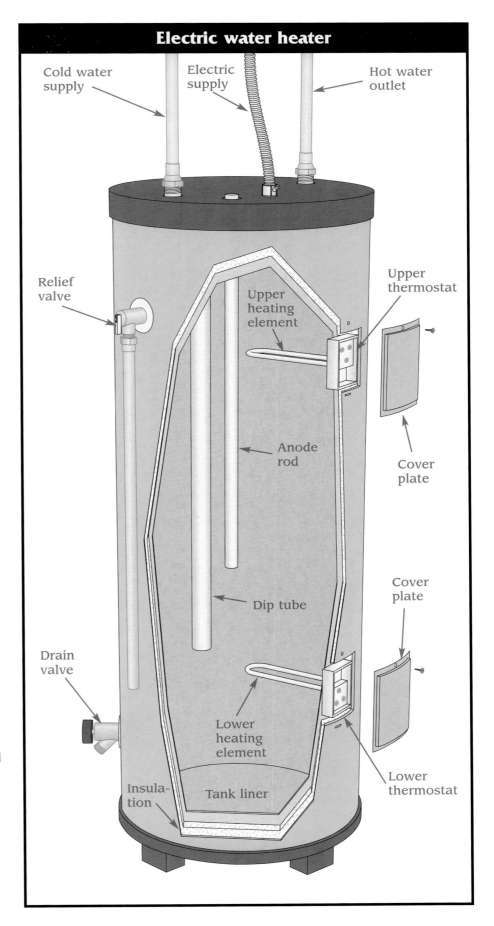

Electric water heater

Cold water supply

Electric supply

Hot water outlet

Relief valve

Upper heating element

Upper thermostat

Cover plate

Anode rod

Dip tube

Cover plate

Drain valve

Lower heating element

Lower thermostat

Insula-tion

Tank liner

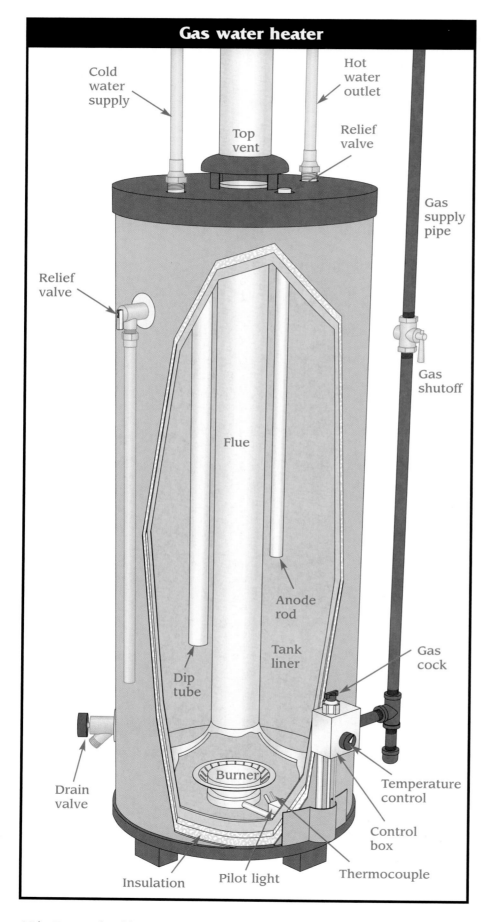

Gas water heater

Cold water supply

Top vent

Hot water outlet

Relief valve

Gas supply pipe

Relief valve

Gas shutoff

Flue

Anode rod

Dip tube

Tank liner

Gas cock

Drain valve

Burner

Temperature control

Insulation

Pilot light

Control box

Thermocouple

Follow the installation instructions carefully, and don't hesitate to call in a professional installer.

Choosing a water heater. When choosing a new heater, try to match your hot water needs with the tank capacity, and save yourself money over the long haul by choosing an energy efficient model with a good warranty. A 40-gallon tank is needed for most households, but if you pride yourself on water-use efficiency, a smaller tank will use less energy. Gas heaters cost significantly less to operate than electric heaters. Spend a little extra on a well-insulated tank too. Energy guide stickers on the appliances tell you how much the heater costs to run, relative to other heaters. If you use the sticker to balance the savings of additional insulation versus initial cost, remember that energy costs may go up. Finally, make sure you have enough space for a new water heater. Measure the space in which you want to put it. The nameplate on the front of the heater will give required clearances for the top and all sides of the appliance. The nameplate will also tell you what kind of gas the appliance uses. Only operate a gas heater on the kind of gas specified on the nameplate. If you have very hard water, consider investing in a water softening system or service. Hard water shortens the life of water heaters and can decrease their energy efficiency by 20 to 30%. Purchasing a water heater with an extra anode rod provides additional corrosion resistance, if this has been a problem with your pipes and heaters. Special aluminum anodes may be needed if your hot water produces odors.

Buy quality copper or stainless steel connectors. The fittings should be solid brass (not brass-plated steel) and the spiral corrugations on flexible connectors should be rounded and spaced, not sharp and tight fitting.

Scale build-up caused by hard water takes a toll on plumbing systems and appliances, like washing machines, that tap into the water supply. A whole-house water conditioning/softening system (called an ion exchanger) greatly reduces scaling.

Water conditioning & purification systems

There may be something you just don't like about your water. Perhaps in smells or tastes bad, leaves iron stains on fixtures, or doesn't get clothes clean. Or, maybe you smell chemicals or excessive chlorine in the shower. It's time to do a little investigating.

If you have municipal water, you can get a copy of the "Municipal Drinking Water Contaminant Analysis Report" from your water utility. Go back two or three years to see if there have been any violations. Also, find out from your health department if lead is a problem in water in your area. Lead leaches out of pipes after it leaves the treatment plant, so it won't show up in the water utility analysis.

If you have a private well, you can find out about common water contaminants in your area from the local health department. Let them know if you live near a dump, a gas station, a feedlot, or some other potential source of water contamination. They can tell you what to test for and steer you to a certified testing laboratory in your area, or they may provide free testing themselves. According to the EPA, residential wells should at least be tested for nitrates and coliform bacteria yearly. Test more frequently and for other contaminants if a problem is suspected. The cost of a test depends on which and how many contaminants the lab tests for, so try to narrow down your search. The lab or your utility will probably compare contaminant levels in your water with EPA established Maximum Contaminant Levels (MCLs); otherwise, you can get MCL tables from the EPA. These tables tell you which (and at what levels) chemicals are dangerous, or which

A Reverse Osmosis Unit (RO) removes most impurities water distillers remove. They will also remove Atrazine, which distillers don't remove. They are generally used to treat cooking and drinking water. Point-of-use RO units, like the one above which mounts in the basement or beneath a sink, clear a greater quantity of water much more quickly and with much less energy than point-of-use distillation units.

just smell, taste, or clean poorly or damage plumbing and appliances.

Solution: A water filtration system

Once you determine the problem areas of your water, you can search for a treatment system. Very turbid (cloudy) water can be cleared up with an activated granular carbon filter with an added sediment filter. Hydrogen sulfide ("rotten egg smell"), high chlorine, trihalomethanes (chemicals formed when chlorine reacts with organic matter) and some other potentially health-threatening organic compounds and pesticides can be removed with a carbon filter. Backwashable carbon filters last longer than filters that can't be back washed, especially for water high in sediments. Large carbon filters last longer and remove more contami-

Fixed-spout water filters contain replaceable filtration cartridges that improve the taste and general quality of drinking and cooking water, but remove less than larger filters and don't last as long.

nants than small filters. Filters installed for the whole house are called point of entry (POE) filters. Filters for just one sink or shower are called point of use (POU).

If you have high levels of total dissolved solids (more than 250 parts per million), you'll need a distillation or reverse osmosis (RO) unit in the kitchen for drinking, cooking and ice-making. These devices also take out dangerous inorganic materials which carbon usually misses—like lead, sodium, arsenic and nitrate. Most RO units include activated carbon filters to take care of the organic compounds and radon that RO alone misses. Distillers remove more total dissolved solids than RO units, but RO units filter more water, more quickly and with much less energy, making them the more popular choice.

If you have hard water or other mineral problems, you'll need an ion exchanger (water softener). Hard water leaves white laundry gray, and encrusts water heaters and pipes with scale, drastically reducing their energy efficiency and life. Ion exchangers are the cheapest, most efficient way to remove hardness minerals like calcium, magnesium, and dissolved iron (undissolved iron is visible in the water and can be removed with a backwashable sediment filter). Ion exchangers can also be set up to remove nitrate, arsenic, chromium, radium, fluoride and uranium. Ion exchangers add small amounts of sodium to water. Potassium salts can be used in the brine tank if you don't want extra sodium in your water.

In wells, most bacteria can be controlled with chlorine or some other disinfectant. Some whole-house carbon filters include ultraviolet (UV) light chambers to destroy organisms without the use of chemicals. POU distillation units also remove bacteria. Cysts, which are a dormant stage of protozoan parasites like Giardia and Cryptosporidium, are not easily destroyed by chlorine or UV. Distillation, RO, and a few carbon and fine-sediment filters can remove cysts.

Important: Water treatment units need to be maintained or they can actually make your water worse. Mark your calendar for when you have to change a membrane or carbon cartridge.

Shower water.

Most water contaminants are worse to drink than to bathe in; however, your absorption of chlorine, volatile organic compounds (called VOCs) and radon are higher in a hot shower than if you drink them. The hot-water spray of water in the shower causes much of the volatile compounds or elements to vaporize, and you breath the vapor into your lungs. Showerhead filters can remove chlorine, but only a whole house filter with activated carbon can remove VOCs and radon.

Chlorine: Too much of a good thing.

Chlorine at high levels may pose health threats. Chlorine reacts with organic matter to produce trihalomethanes like chloroform, a VOC that can cause cancer and miscarriages. If you have a well with high levels of organic matter that also requires heavy doses of chlorine to keep sterile, consider drilling a deeper well or investing in a whole house filter that takes out trihalomethanes. If you're on municipal water, you can request information on the chemistry of your water from your water utility.

Evaluating treatment system claims.

First, read and compare the fine print to see what contaminants and what quantities of contaminants the manufacturers are claiming to remove. Make sure the National Sanitation Foundation (NSF) or the Water Quality Association (WQA) backs up the device and its claims. These organizations can also help you choose the right filter. The NSF can be reached at (800) 673-8010 www.nsf.org. The WQA provides free literature on choosing the right filter. They can be reached at (800) 749-0234 or www.wqa.org.

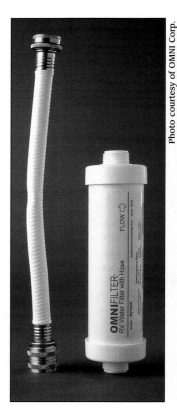

On-demand water filtration units can be used to provide cleaner, tastier water outside of your home plumbing system. The unit shown here is designed for installation in recreational vehicles and boats to remove rust and sediment in these independent water supply systems.

Major Home Water Treatment Technologies

Technology	Effectively removes	Generally won't remove	Comments
Activated Granular Carbon Filters	Some pesticides, mercury, chlorine, trihalomethanes, radon, hydrogen sulfide "rotten egg" taste and smell, other bad tastes, turbidity (cloudiness), reddish color, and iron staining.	Arsenic, asbestos, benzene, fluoride, lead, nitrate, radium, coliform bacteria, cysts (Cryptosporidium, Giardia), metallic taste, soap residue, scale buildup in water heaters and plumbing.	Can be used in POE systems when radon or organic chemicals pose health risk. Usable in POU systems for drinking and cooking if raw water is not too bad to begin with. Can be used with RO units and other materials to remove greater quantities and a broader range of contaminants.
Activated Solid-Block or PreCoat Carbon Filters	Removes everything granular filters remove except radon and turbidity. Also removes asbestos, benzene, lead, and cysts, which granular filters cannot remove.	Arsenic, fluoride, nitrate, radium, radon, bacteria, metallic taste, turbidity, soap residue, scale buildup in water heaters and plumbing.	Can be used in POE or POU systems. May be used with RO units and sediment filters to remove greater quantities and a broader range of contaminants.
Ion Exchangers (Softeners)	Typically used to remove the hard water minerals calcium, magnesium and iron. Can be set up to remove nitrates, arsenic, chromium, fluoride, radium, and uranium for the whole house and less expensively than an RO system.	Does not remove turbidity, organic chemicals, radon, mercury, bacteria, cysts, tastes, or smells. Cannot remove both cations (such as calcium) and anions (such as nitrate) at the same time.	Ion Exchangers are usually used to treat hard water. Installing a different resin material allows anions like nitrate and fluoride to be removed. Some houses use two softener tanks to remove both cations and anions.
Distillers	Removes everything except Radon and VOC's. Not designed to treat turbid water unless equipped with a carbon filter. Generally not used to treat anything except cooking and drinking water.	Atrazine, Benzene, and Radon are not removed. These chemicals and turbidity may be removed by carbon filters built into the distillation unit.	Distillers are POU systems for thoroughly cleaning small batches of water that contain chemical and biological contaminants. May be used in conjunction with a carbon filter to also remove radon and VOCs.
Reverse Osmosis (RO) Units	Removes most things distillers remove but not trihalomethanes and bacteria. Removes Atrazine, which distillers don't remove. Generally used to treat cooking and drinking water.	Trichloromethanes, Benzene and Radon not removed. Carbon filters built into the RO unit may remove these chemicals and turbidity.	RO POU units remove slightly less total dissolved solid material than distillers. But they clear a greater quantity of water much more quickly and with much less energy than distillation units.
Filter media used on hot water at the shower head.	Used to remove sediment, chlorine and a few other substances from hot water at the shower head	Does not remove most organic and inorganic chemicals. Does not remove trihalomethanes, the class of dangerous chemicals that can be produced when organic matter in water reacts with chlorine.	Shower head filters remove chlorine but don't remove trihalomethanes, benzene, radon or other chemicals which can be dangerous to shower in.

Index